Construction Science and Materials

Construction Science and Materials

Second Edition

Surinder Singh Virdi
Lecturer in Construction
South and City College Birmingham

Visiting lecturer
City of Wolverhampton College

With contribution from
Robert Waters
Lecturer in Construction
South and City College Birmingham

WILEY Blackwell

This edition first published 2017
© 2017 by John Wiley and Sons Ltd

Registered Office
John Wiley & Sons Ltd, The Atrium, Southern Gate, Chichester, West Sussex, PO19 8SQ, United Kingdom.

Editorial Offices
9600 Garsington Road, Oxford, OX4 2DQ, United Kingdom.
The Atrium, Southern Gate, Chichester, West Sussex, PO19 8SQ, United Kingdom.

For details of our global editorial offices, for customer services and for information about how to apply for permission to reuse the copyright material in this book please see our website at www.wiley.com/wiley-blackwell.

Library of Congress Cataloging-in-Publication data applied for

ISBN: **9781119245056**

A catalogue record for this book is available from the British Library.

Wiley also publishes its books in a variety of electronic formats. Some content that appears in print may not be available in electronic books.

Cover image: MACIEJ NOSKOWSKI/Gettyimages

Set in 10/12pt Warnock by SPi Global, Pondicherry, India
Printed and bound in Malaysia by Vivar Printing Sdn Bhd

10 9 8 7 6 5 4 3 2 1

How to Use This Book

All students should spend some time studying the first seven chapters.

Students pursuing Level 2 courses should focus additionally on Chapters 9, 10 and 16.

Students pursuing Level 3/4/5 courses should study all chapters in this book.

Specimen assignment tasks are given in Chapter 17, which the students can try once they have studied the relevant topics. The model answers are given on the companion website http://www.wiley.com/go/virdiconstructionscience2e.

The website also includes solutions for the end-of-chapter exercises, information on the use of a scientific calculator, information on units, information on settlement and consolidation, the design of building foundations, the design of timber joists, daylight calculations and PowerPoint presentations on some topics.

Contents

Preface to the Second Edition

This book has been written for students pursuing full-time/part-time studies in level 2, 3, 4 and 5 programmes in Construction, Civil Engineering and Building Services. The book should also be informative for students on level 2/3 construction craft courses. The topics included cover most of the syllabus of the core subject of Construction Science and Materials. The syllabi cover a wide range of topics, and since Construction Materials is a subject on its own, the discussion in this book is focussed on a selection of nine materials that are used widely in building and civil engineering projects. Structural Mechanics is complex and is also a subject on its own; I have tried to include information on some of the basic concepts that students need to learn to achieve the relevant grading criteria.

The learning material has been divided among the first sixteen chapters, which provide information on construction science, construction materials and structural mechanics for the above courses. Two chapters from the first edition have been moved to the companion website, and two new chapters – Chapter 6 (Introduction to building services) and Chapter 11 (Forces and structures 3) – have been included in the second edition. Each chapter gives detailed explanations of the topics involved, and the text in the second edition is supported by a large number of illustrations and worked examples. To reinforce students' learning, almost all chapters have end-of-chapter exercises, and if a student has difficulty in obtaining the right answer, help is at hand in the form of solutions available on the companion website.

My thanks are due to my family, my students and colleagues for the interest they have shown in this project, and a special thank you to Robert Waters for his contribution towards the development of new material for this edition.

A big thank you to: Madeleine Metcalfe, Viktoria Vida (Editorial Assistant), Blesy Regulas (Project Editor) and Rajitha Selvarajan (Production Editor) for their support during the publication of this book.

Surinder Singh Virdi

About the Companion Website

Don't forget to visit the companion website for this book:

http://www.wiley.com/go/virdiconstructionscience2e

There you will find valuable material designed to enhance your learning, including:

1) Fully worked solutions to the exercises at the ends of chapters;
2) Model answers for the assignment tasks set in Chapter 17;
3) Explanations of settlement and consolidation in structures; details on the design of building foundations; and daylight calculations;
4) A task + solution on the design of timber joists;
5) PowerPoint slides for lecturers on: Hooke's Law; Forces and their Effects; Temperature and Heat Loss.

Scan this QR code to visit the companion website:

1

Introduction to Physics

LEARNING OUTCOMES
1) Define speed, velocity and acceleration.
2) Explain mass, gravitation and weight.
3) Explain Newton's laws of motion and solve numerical problems based on these laws.
4) Explain work, energy and power, and solve numerical problems.

1.1 Speed and Velocity

In the study of moving objects, one of the important things to know is the rate of motion. The rate of motion of a moving object is what we call **speed**. It may be defined as the distance covered in a given time:

$$\text{Speed} = \frac{\text{Distance covered}}{\text{Time taken}}$$

If the distance covered is in metres (m) and the time taken in seconds (s), then speed is measured in metres per second (m/s). If the distance is in kilometres (km) and the time in hours (h), the unit of speed is kilometres per hour (km/h).

When the direction of movement is combined with the speed, we have the **velocity** of motion. Quantities that have both magnitude and direction are known as **vector** quantities. Velocity is a vector quantity; its magnitude and direction can be represented by an arrow. Speed, on the other hand, has magnitude but no direction; therefore it is called a **scalar** quantity.

1.2 Acceleration

An object is said to accelerate if its velocity increases. The rate of increase of velocity is called the **acceleration**.

$$\text{Acceleration} = \frac{\text{Increase in velocity}}{\text{Time taken}}$$

Construction Science and Materials, Second Edition. Surinder Singh Virdi.
© 2017 John Wiley & Sons Ltd. Published 2017 by John Wiley & Sons Ltd.
Companion website: www.wiley.com/go/virdiconstructionscience2e

If velocity is measured in metres and time in seconds, then acceleration is measured in metres per second per second (m/s/s) or metres per second squared (m/s²). If the velocity of a moving object decreases, it is said to decelerate, i.e. the acceleration is negative. The following relationships may be used to solve problems involving velocity and acceleration:

- $v^2 - u^2 = 2as$
- $v = u + at$
- $v = ut + \dfrac{1}{2}at^2$

where, u = initial velocity
   v = final velocity
   a = acceleration
   t = time
   s = distance

1.3   Mass

The amount of matter contained in an object is known as its **mass**. The basic SI unit of mass is the kilogram (kg).

   $1 \text{ gram}(g) = 1000 \text{ milligrams}(mg)$
   $1000 \text{ grams} = 1 \text{ kilogram}$
   $1000 \text{ kilograms} = 1 \text{ tonne}(t)$

The mass of an object remains constant irrespective of wherever it is.

1.4   Gravitation

Gravitation can be defined as the force of attraction that exists between all objects in the universe. According to Isaac Newton, every object in the universe attracts every other object with a force directed along the line of centres for the two objects that is proportional to the product of their masses and inversely proportional to the square of the distance between their centres.

$$F_g = \frac{G\, m_1\, m_2}{r^2}$$

where F_g = gravitational force between two objects
   m_1 = mass of first object
   m_2 = mass of second object
    r = distance between the centres of the two objects
   G = universal constant of gravitation

The value of constant G is so small that the force of attraction between any two objects is negligible. In 1798, Henry Cavendish performed experiments to determine the value of G and found it to be 6.67×10^{-11} Nm²/kg².

If we consider an object and the Earth, the mass of Earth is so large (5.98×10^{24} kg) that, depending on the mass of the object, there could be a considerable force of attraction between the two. That is why when an object is dropped from a height, it falls towards the Earth, not away from it. The initial velocity of the object is zero m/s, but as the distance increases, the velocity of the falling object also increases. The rate of increase in velocity is called acceleration and, in the case of a free-falling object, it is known as the **acceleration due to gravity** (symbol: g).

The value of g is 9.807 m/s^2, but for all calculations in this book it will be approximated to 9.81 m/s^2 (m/s^2 can also be written as ms^{-2}).

Example 1.1   Find the gravitational force between the Earth and:

a) An object with a mass of 1 kg.
b) A person with a mass of 80 kg.

Given: mass of the Earth = 6.0×10^{24} kg; radius of the Earth = 6.4×10^6 m; $G = 6.7 \times 10^{-11}$ Nm^2/kg^2

Solution:

a)
$$F_g = \frac{G\, m_1\, m_2}{r^2}$$
$$= \frac{6.7 \times 10^{-11} \times 6 \times 10^{24} \times 1}{\left(6.4 \times 10^6\right)^2}$$
$$= 9.81 \text{ N}$$

b)
$$F_g = \frac{G\, m_1\, m_2}{r^2}$$
$$= \frac{6.7 \times 10^{-11} \times 6 \times 10^{24} \times 80}{\left(6.4 \times 10^6\right)^2}$$
$$= 785.16 \text{ N}$$

1.5 Weight

The **weight** of an object is the force with which it is attracted towards Earth. When an object falls freely towards Earth, the average value of the acceleration produced (g) is 9.81 m/s^2. The force (F) acting on the object due to Earth's gravitational pull (or the weight of the object) can be calculated as:

$$F = m \times g$$

where m is the mass of the object in kg.

The units of weight are the same as the units of force. If the mass is in kilograms, the unit of weight will be newtons (N).

The weight of a 1 kg mass will be:

$$F = 1 \times 9.81 = 9.81 \text{ N}$$

Similarly, the weight of a 5 kg mass is:

$$F = 5 \times 9.81 = 49.05 \text{ N}$$

For larger forces, kilonewtons or meganewtons may be used.

$$1000 \text{ N} = 1 \text{ kilonewton}(\text{kN})$$
$$1000000 \text{ N} = 1 \text{ meganewton}(\text{MN})$$

The weight of a body is not constant but changes slightly when we move from the Equator to the North Pole. The Earth is not a perfect sphere: it bulges at the Equator. This affects the gravitational force, which varies from 9.78 m/s^2 at the Equator to 9.83 m/s^2 at the North Pole.

1.6 Volume

All substances, whether they are solid, liquid or gas, occupy space. The amount of space occupied by an object is called its **volume**.

$$\text{Volume} = \text{length} \times \text{width} \times \text{height}$$
$$= \text{area} \times \text{height}$$
$$(\text{Units: m}^3, \text{ cm}^3 \text{ or mm}^3.)$$

1.7 Density

If equal volumes of bricks, concrete, timber and other materials are compared, the values of their mass will be different. This is because different materials do not have the same density.

The **density** of a material is defined as its mass per unit volume.

$$\text{Density} = \frac{\text{Mass}}{\text{Volume}}$$

If the units of mass and volume are kg and m^3 respectively, then the unit of density will be kilograms per metre cubed (kg/m^3). The density of pure water is 1000 kg/m^3. The density of a material is an important property and is used in several areas of building technology, for example:

1) To find the self-weight (dead load) of a component like a beam, column etc., its density must be known.
2) The strength of a material, generally, depends on its density.
3) The thermal insulation of a material is inversely proportional to its density.

Table 1.1 shows the densities of a selection of materials.

Table 1.1

Material	Density (kg/m³)
Concrete blocks (lightweight)	450–675
Aluminium	2720
Brick (common)	2000
Brick (engineering)	2200
Cement	1500
Concrete	2400
Copper	8800
Cork	200
Glass	2500
Granite	2720
Gravel (coarse)	1450
Gravel (all-in)	1750
Lead	11300
Limestone	2250
Marble	2720
Mercury	13500
Mild steel	7820
Sand (dry)	1600
Sandstone	2250
Slate	2800
Timber (Oak)	600–900
Timber (Beech)	700–900

Example 1.2 The mass of a concrete block measuring $250\,\text{mm} \times 200\,\text{mm} \times 200\,\text{mm}$ is $24.0\,\text{kg}$. Find the density of concrete.

Solution:
The dimensions of the concrete block are converted into metres to obtain the density in kg/m^3.

$$250 \text{ mm} = \frac{250}{1000}\text{m} = 0.250 \text{ m}$$

Similarly, $200\,\text{mm} = 0.200\,\text{m}$

$$\text{Volume of the concrete block} = 0.250 \times 0.200 \times 0.200 = 0.010 \text{ m}^3$$
$$\text{Density} = \frac{\text{Mass}}{\text{Volume}}$$
$$= \frac{24}{0.010} = 2400 \text{ kg/m}^3$$

Example 1.3 The cross-sectional measurements of a 7.0 m long concrete beam are 0.3 m × 0.75 m. Find the mass and the weight of the beam. Density of concrete = 2400 kg/m³.

Solution:

$$\text{Volume of the beam} = 7.0 \times 0.3 \times 0.75 = 1.575 \text{ m}^3$$
$$\text{Mass} = \text{Density} \times \text{Volume}$$
$$= 2400 \times 1.575 = 3780 \text{ kg}$$
$$\text{Weight} = \text{Mass} \times g$$
$$= 3780 \times 9.81 = 37081.8 \text{ N}$$

1.8   Specific Gravity

The specific gravity of a substance is defined as the ratio of the density of the material to the density of water.

$$\text{Specific gravity} = \text{Density of a material} \div \text{Density of water}$$

The specific gravity of a material remains the same, irrespective of the units of density.

1.9   Newton's First Law of Motion

In the seventeenth century, Isaac Newton formulated three laws, which are known as Newton's laws of motion. The first law states that an object will remain in a state of rest or uniform motion in a straight line unless acted upon by an external force. This means that a book lying on a desk will lie there forever unless somebody applies an effort (external force) to pick it up. Similarly, imagine you are travelling in a car at, say, 60 km/hr and the ignition is turned off. The car will eventually come to a halt without the application of brakes. This is due to the friction between the car tyres and the road surface. Friction is a force that tries to stop moving objects. The car will not stop if there is no friction between the car tyres and the road surface. In space there is no influence of external forces. A spacecraft will continue to travel in a straight line at a constant speed. It does not need a force to keep it moving.

1.10   Newton's Second Law of Motion

Newton's second law of motion states that when an unbalanced force acts on an object, the object will accelerate in the direction of the force. The acceleration is directly proportional to the force and inversely proportional to the mass.

$$a = \frac{F}{m} \quad \text{or,} \quad F = ma$$

where   F = the force (newtons)
m = the mass of the object (kg)
a = the acceleration produced (m/s²)

A force of 1 newton gives a mass of 1 kg an acceleration of 1 m/s²

Example 1.4   Calculate the acceleration of a 100 kg object if it is acted upon by a net force of 250 N.

Solution:

$$F = ma \text{ (a is the acceleration)}$$
$$250 = 100 \times a$$
$$\frac{250}{100} = a \text{ or, } a = 2.5\,\text{m/s}^2$$

1.11   Newton's Third Law of Motion

Newton's third law of motion states that to every action there is an equal and opposite reaction. Consider a beam resting on two walls, as shown in Figure 1.1.

The weight of the beam plus any other force is the action. The reactions (R_1 and R_2) are offered by the walls as they support the beam and resist its downward movement. For the stability of the beam, the total reaction must be equal to the action.

$$\text{Weight of the beam} + \text{Forces acting on the beam} = R_1 + R_2$$

If the walls cannot support the beam, due either to some defect in the wall or to the use of weaker materials, the reaction will not be equal to the action and the beam will move away from its intended position. Depending on the magnitude of the movement, this might cause the failure of the component that the beam is supporting.

1.12   Friction

When an object rests on a surface, two forces act on it to maintain the balance: the weight acting downwards and the reaction (or normal reaction) acting upwards, as shown in Figure 1.2. If a force F is applied to slide the object, the movement is resisted by another force that acts in the opposite direction, as shown in Figure 1.3. The opposing force is called the **friction force** (R) and is due to the roughness of the surfaces in contact.

If the applied force is increased, the friction force increases as well. The maximal friction force is experienced when the object is about to move. This is called **static friction**. Friction also acts when the object is in motion, but this type of friction (called **dynamic friction**) is less than static friction.

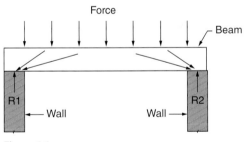

Figure 1.1

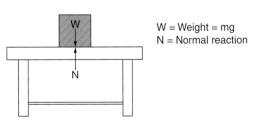

W = Weight = mg
N = Normal reaction

Figure 1.2

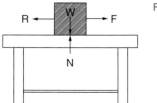

Figure 1.3

The amount of friction between two surfaces depends on:

1) The normal reaction, which acts at right angles to the two surfaces.
2) The roughness of the surfaces in contact.

The coefficient of friction (μ) is given by:

$$\mu = \frac{\text{Friction force}\,(R)}{\text{Normal reaction between surfaces}\,(N)}$$

$$= \frac{R}{N} = \frac{F}{W} \quad (R = F;\ N = W)$$

Example 1.5   A horizontal force of 9.0 N moves a brick on a metal surface at a uniform speed. Find the weight of the brick if the coefficient of friction between the two materials is 0.45.

Solution:

$$R = F = 9.0\ \text{N};\ N = W;$$

$$\mu = \frac{R}{N} = \frac{R}{W}$$

$$W = \frac{R}{\mu} = \frac{9.0\text{N}}{0.45} = 20.0\ \text{N}$$

1.13   Work

Work is said to be done when a force moves an object. The work done can be calculated from the following equation:

$$\text{Work done} = \text{Force} \times \text{Distance moved}$$

or, $W = F \times s$

The SI unit of work is the **joule** (J), which can be defined as the work done when a force of 1 newton moves through a distance of 1 m in the direction of the force.

Example 1.6   A 50 cm × 50 cm × 50 cm block of concrete rests on a concrete floor. The coefficient of friction between the two surfaces is 0.6. Calculate:

a) The horizontal force necessary to move the concrete block.
b) The work done in moving the block by 10 m.

The density of concrete is 2400 kg/m^3

Solution:

a) Mass $= $ Density $\times$ Volume

$$= 2400 \times (0.5 \times 0.5 \times 0.5)(50 \text{ cm} = 0.5 \text{ m})$$
$$= 300 \text{ kg}$$

Weight of the block $(W) = 300 \times 9.81 = 2943 \text{ N}$

Coefficient of friction, $\mu = \dfrac{F}{W}(F$ is the horizontal force$)$

$$0.6 = \dfrac{F}{2943}$$
$$F = 0.6 \times 2943 = 1765.8 \text{ N}$$

b) Work done $=$ Force $\times$ Distance

$$= 1765.8 \times 10$$
$$= 17658 \text{ J or } 17.658 \text{ kJ}$$

1.14 Energy

The capacity to do work is known as **energy**. Energy may be available in various forms but it is not possible to create or destroy energy. However, it may change from one form to another, for example, from light energy into electrical energy, from electrical energy into heat energy, from heat energy into electrical energy etc.

Some of the main forms in which energy exists are:

- Chemical energy;
- Electrical energy;
- Kinetic energy;
- Light energy;
- Nuclear energy;
- Potential energy;
- Sound energy;
- Thermal energy.

Potential energy and kinetic energy are discussed further in the next two sections.

1.14.1 Potential Energy

Potential energy may be defined as the energy possessed by a body due to its position above the ground. If an object of mass m kilograms is raised to a height h metres, then the work done in doing so is given by:

$$\text{Work done} = \text{Force} \times \text{Distance}$$
$$= (m \times g) \times h = mgh$$

The potential energy (PE) possessed by the object, at height h metres, is mgh.
The work done by the object, if allowed to fall, is also mgh.
The unit of energy is the joule (J).

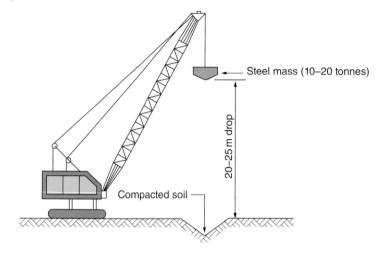

Figure 1.4 Dynamic compaction.

1.14.2 Kinetic Energy

The energy possessed by a moving object is known as the **kinetic energy** (KE).

$$\text{Kinetic energy} = \frac{1}{2}mv^2$$

where m = the mass of the object in kg
     v = the velocity of the object in m/s

There are several uses of potential energy and kinetic energy in civil engineering, two of which are: hydroelectric power stations and the improvement of loose subsoil.

In hydroelectric power stations, water is stored in the form of a lake by constructing a concrete dam or an earth dam. The water level rises and, due to its height, possesses energy. The water is allowed to fall through a pipe (penstock) and its energy is used to drive a turbine. The turbine, in turn, generates electricity.

Loose subsoils are not very strong and hence may not be able to support a building/structure satisfactorily. The strength of the subsoils may be improved by several techniques, one of which is called **dynamic compaction**. The method involves dropping a heavy block of steel from a suitable height (Figure 1.4). As the block falls on the ground, its energy is used to compact the soil. The compaction of a soil results in the improvement of its density and strength.

Example 1.7   A 10 tonne block of steel was raised to a height of 12.0 m and then dropped. Calculate the energy possessed by the block at heights of 12 m, 9 m, 6 m, 3 m and when it hit the ground.

Solution:

10 tonnes = 10000 kg

The steel block is at rest at a height of 12 m, i.e. its velocity, and hence the kinetic energy, is zero. The energy possessed by it is entirely due to its height above the ground surface.

Total energy at 12 m height = Potential energy = mgh
$$= 10000 \times 9.81 \times 12.0 = 1177200 \text{ J}$$

Table 1.2

Height (m)	$v^2 = 2as + u^2$	Kinetic energy $= \frac{1}{2}mv^2$ (J)	Potential energy $=$ mgh (J)	Total energy $=$ P E $+$ K E (J)
12	0	0	1177200	1177200
9	58.86	294300	882900	1177200
6	117.72	588600	588600	1177200
3	176.58	882900	294300	1177200
0	235.44	1177200	0	1177200

At a height of 9 m: As the block is dropped, it accelerates due to gravitational force. The energy possessed by the block is a combination of kinetic energy and potential energy. Its velocity, v, can be determined from:

$$v^2 - u^2 = 2as \text{ or, } v^2 = u^2 + 2as \quad (\text{initialvelocity, } u = 0)$$

$$v^2 = 0^2 + 2 \times 9.81 \times 3.0 = 58.86 \quad (s = 12 - 9 = 3 \text{ m})$$

$$\text{Kinetic energy} = \frac{1}{2}mv^2$$

$$= \frac{1}{2} \times 10000 \times 58.86 = 294300 \text{ J}$$

$$\text{Potential energy} = \text{mgh}$$

$$= 10000 \times 9.81 \times 9.0 = 882900 \text{ J}$$

$$\text{Total energy at a height of 9.0 m} = \text{KE} + \text{PE}$$

$$= 294300 + 882900 = 1177200 \text{ J}$$

The calculations at the other heights can be made in a similar way and are summarised in Table 1.2.

These calculations show that the total energy possessed by the steel block at the 12 m height is entirely due to the potential energy. The total energy is entirely due to its kinetic energy when the block hits the ground, i.e. at zero height. At other heights, the total energy of the block is a combination of kinetic and potential energies.

1.15 Power

Power is defined as the work done per second, i.e. the rate of doing work.

$$\text{Power} = \frac{\text{Work done}}{\text{Time taken}}$$

Power is measured in joules per second (J/s) or watts (W).

$$\text{One watt} = \text{one joule of work done per second}$$
$$\text{or, } 1 \text{ W} = 1 \text{ J/s}$$

$$\text{also, Power} = \frac{\text{Work done}}{\text{Time taken}} = \frac{\text{F} \times \text{s}}{\text{t}}$$

$$= \text{Force} \times \text{Velocity} \left(\frac{\text{s}}{\text{t}} = \text{velocity} \right)$$

Example 1.8   A crane lifts a block of concrete with a mass of 2.0 tonnes to a height of 8.0 m in 12 seconds. Find the work done in lifting the block and the power output of the crane.

Solution:

$$2.0 \text{ tonnes} = 2000 \text{ kg}$$
$$\text{Work done} = \text{Force} \times \text{Distance}$$
$$= (2000 \times 9.81) \times 8.0$$
$$= 156960 \text{ J}$$
$$\text{Power} = \frac{156960}{12} = 13080 \text{ W}$$

Exercise 1.1

1   Find the gravitational force between the Earth and:
  A An object with a mass of 10 kg.
  B A person with a mass of 60 kg.
  Given: mass of the Earth $= 6.0 \times 10^{24}$ kg; radius of the Earth $= 6.4 \times 10^{6}$ m; $G = 6.7 \times 10^{-11}$ Nm^2/kg^2.

2   A block measuring 300 mm × 200 mm × 200 mm is made from concrete with a density of 2300 kg/m^3. Find the mass of the block.

3   The cross-sectional measurements of a 6.0 m long concrete beam are 0.3 m × 0.7 m. Find the mass and the weight of the beam. The density of concrete is 2400 kg/m^3.

4   Calculate the acceleration of an 80 kg object if it is acted upon by a net force of 960 N.

5   A horizontal force of 85.0 N moves a concrete component on a metal surface at a uniform speed. Find the weight of the component if the coefficient of friction is 0.45.

6   An 80 cm × 50 cm × 50 cm block of concrete rests on a concrete floor. The coefficient of friction between the two surfaces is 0.6. Calculate:
  A The horizontal force necessary to move the concrete block.
  B The work done in moving the block by 15 m.
  The density of concrete is 2300 kg/m^3

7   An 8 tonne block of steel is raised to a height of 9.0 m and then dropped. Calculate the energy possessed by the block at heights of 9 m, 6 m, 3 m and when it hits the ground.

8   A crane lifts a block of concrete with a mass of 2.5 tonnes to a height of 10.0 m in 12 seconds. Find:
  A The work done in lifting the block.
  B The power output of the crane.

Reference/Further Reading

1 Breithaupt, J. (2008). *AQA Physics A*. Cheltenham: Nelson Thornes.

2

Introduction to Chemistry

LEARNING OUTCOMES

1) Define atom, electron, proton and neutron.
2) Explain electrovalency and covalency.
3) Explain the difference between elements, compounds and mixtures.
4) List a range of symbols and formulae.
5) Differentiate between acids and bases.

2.1   Introduction

The concept that matter is composed of tiny particles has been known to scientists for many centuries. John Dalton, a British scientist, put forward his atomic theory in 1808 and suggested that matter is composed of minute particles called atoms, which cannot be created, destroyed or split. Another scientist, Marie Curie, and her husband Pierre discovered two new elements called polonium and radium. The salts of these elements glowed in the dark and Marie Curie called this property **radioactivity**.

An **atom** is the smallest particle that can exist on its own, and is made up of three types of particles:

- Electrons
- Protons
- Neutrons

The electrons carry a negative charge, the protons a positive charge and the neutrons, as the name suggests, do not carry any charge. Many scientists, during the last century and earlier, assumed that in an atom, protons, electrons and neutrons were mixed up. Lord Rutherford, a British physicist, suggested that the protons and neutrons formed the nucleus of an atom, with electrons surrounding the nucleus, as shown in Figure 2.1.

Figure 2.1 is a highly enlarged representation of an atom. Atoms can be of different sizes, the diameter of an average atom being 0.5×10^{-9} m. The mass of a proton and a neutron is the same as that of an atom of hydrogen. This is known as 1 atomic mass unit and is equal to 1.67×10^{-27} kg. The mass of an atom depends on the number of protons and neutrons. The mass of electrons is very small, as compared to that of protons and neutrons; hence, they do not contribute to the mass of an atom.

Construction Science and Materials, Second Edition. Surinder Singh Virdi.
© 2017 John Wiley & Sons Ltd. Published 2017 by John Wiley & Sons Ltd.
Companion website: www.wiley.com/go/virdiconstructionscience2e

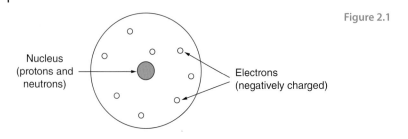

Figure 2.1

Table 2.1 Atomic numbers of some elements.

Element	Atomic number	Number of protons	Number of electrons
Oxygen	8	8	8
Aluminium	13	13	13
Calcium	20	20	20
Copper	29	29	29
Iron	26	26	26

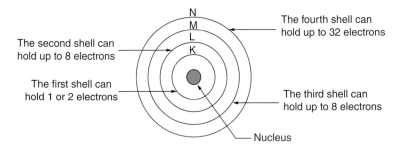

Figure 2.2

The electrical charge carried by a proton and an electron is +1 unit and −1 unit, respectively. One unit of charge is equal to the charge on the nucleus of a hydrogen atom. Another important feature of modern atomic theory is that the number of electrons in an atom is the same as the number of protons; this is responsible for the neutral charge on an atom. The number of protons or electrons in the atom of an element is called its **atomic number**. Table 2.1 summarises the atomic numbers of some elements.

The number of protons and neutrons is called the **mass number**. Sodium has 11 protons and 12 neutrons. The mass number is 11 + 12 = 23. The mass number and the atomic number are written above and below the chemical symbol for the element, for example, $^{23}_{11}\text{Na}$.

According to atomic theory, the electrons are in constant motion around the nucleus in paths called **orbits**. As all electrons do not possess the same energy, those having less energy circle the nucleus in orbits with smaller radii whereas the electrons having more energy move around the nucleus in orbits with larger radii.

Orbits of similar energy level can be grouped together, each group being known as a **shell**. There can be several shells around a nucleus, each capable of holding a maximum number of electrons. The shells can be represented by the letters K, L, M etc., as shown in Figure 2.2.

Figure 2.3 Electron configuration of aluminium.

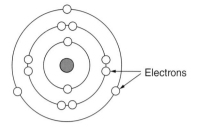

Figure 2.4 Electron configuration of (a) carbon; (b) oxygen; and (c) calcium.

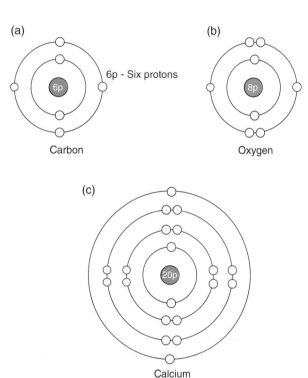

Most of the volume of an atom is space. The protons and neutrons are at the centre of an atom in the part known as the **nucleus**. The electrons are present in the space surrounding the nucleus and are in constant motion. The electron configuration of an atom can be worked out if the atomic number of the element is known. Figure 2.3 shows the electron configuration of aluminium. The atomic number of an atom of aluminium is 13, which means that there are 13 electrons in an atom. The first shell can hold 2 electrons, the second shell 8 and the third shell has 3.

The electron configurations of carbon, oxygen and calcium are shown in Figure 2.4.

2.2   Electrovalency and Covalency

The electron configurations of some of the elements have already been discussed in the previous section. Many elements have incomplete outside shells. When two of these elements are mixed, they react in order to have complete outside shells. Metal atoms (aluminium, copper, iron etc.) lose their extra electron(s), which are gained by the non-metal atoms. In this way, the outside shells of both the metal and the non-metal atoms become complete.

Originally, a metal atom, like all other atoms, is neutral as the number of protons (positive charge) is the same as the number of electrons (negative charge). On losing the spare electrons from the last shell, the metal atom now has more protons than electrons, and hence it carries a positive charge. On the other hand, the atoms of non-metals gain electrons to complete the last shell. This makes them negatively charged as the number of electrons now exceeds the number of protons. Atoms that carry either a negative or a positive charge are known as **ions**. The compounds made by elements by gaining or losing electrons are called **electrovalent compounds**. To understand the phenomenon of electrovalency, the structure of potassium chloride is considered here. Potassium and chlorine combine to give potassium chloride:

Potassium + chlorine = potassium chloride

When potassium and chlorine react, the potassium atom can lose the electron in the last shell. The third shell now has 8 electrons and is, therefore, complete and stable. The electron lost by the potassium atom is gained by the chlorine atom to complete its last shell. Thus, both elements end up with complete outer shells (see Figure 2.5). Potassium becomes a positively charged potassium ion (cation) after losing one electron. Similarly, chlorine gains one electron and becomes a negatively charged chloride ion (anion). The oppositely charged ions attract each other, i.e. cations attract anions, and vice versa. The bond between anions and cations is called an ionic bond or **electrovalent bond**. The number of electrons gained or lost by an atom

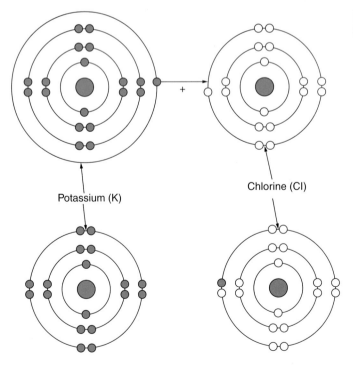

Figure 2.5 Formation of compound potassium chloride.

Potassium (K)

Chlorine (Cl)

when two elements react is known as its electrovalency or, in short, **valency**. The valencies of potassium and chlorine are:

$$\text{Valency of potassium} = 1(\text{positive}); \text{ written as } K^+$$
$$\text{Valency of chlorine} = 1(\text{negative}); \text{ written as } Cl^-$$

2.2.1 Covalent Bond

Some elements do not form ions, but can still combine with other elements making covalent compounds. A chlorine atom has seven electrons in the outermost shell. When two atoms of chlorine combine, both atoms want to gain electrons to complete their outermost shells. This is achieved by the atoms sharing two electrons (Figure 2.6).

One pair of electrons is shared by both atoms of chlorine; therefore, each atom has eight electrons in the outermost shell. A bond between two atoms due to the sharing of electrons is called a **covalent bond**. The valency of the element is the same as the number of pairs of electrons that one atom shares with the other. Water is also a covalent compound. An oxygen atom has six electrons in the outermost shell; a hydrogen atom has one. Two atoms of hydrogen and one atom of oxygen combine, as shown in Figure 2.7. A hydrogen atom shares one pair of electrons, an oxygen atom shares two; therefore, the valency of hydrogen is one and that of oxygen is two. The valencies of ions of some of the metals and non-metals are given in Table 2.2. Copper ions have a positive charge and a valency of 2; this is shown in Table 2.2 as Cu^{2+}. Similarly, chloride ions, which have a negative charge and a valency of one, are shown as Cl^-.

Figure 2.6 Covalent bond.

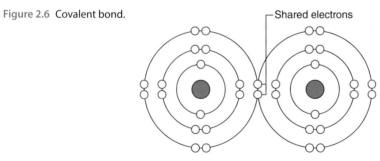

Figure 2.7

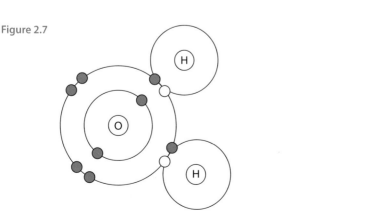

Table 2.2

Ions with positive charge	Valency
Sodium	Na^+
Potassium	K^+
Hydrogen	H^+
Calcium	Ca^{2+}
Magnesium	Mg^{2+}
Copper	Cu^{2+}
Lead	Pb^{2+}
Zinc	Zn^{2+}
Iron (Ferrous)	Fe^{2+}
Iron (Ferric)	Fe^{3+}
Aluminium	Al^{3+}

Ions with negative charge	Valency
Hydroxide ion	OH^-
Nitrate	NO_3^-
Chloride	Cl^-
Carbonate	CO_3^{2-}
Oxide	O^{2-}
Sulphate	SO_4^{2-}

2.3 Elements and Compounds

An **element** is a substance that cannot be decomposed into simpler substances. There are more than 100 elements, some of which are:

Calcium, copper, zinc, aluminium, chlorine, oxygen, carbon, sulphur

In chemistry, each element is given a symbol, which makes representation of chemical reactions very convenient. Table 2.3 shows the symbols of a selection of elements.

$$Hydrogen + oxygen \rightleftharpoons Water$$
$$2H_2 + O_2 \rightleftharpoons 2H_2O$$

Water is produced by combining hydrogen and oxygen gases. Also, by passing current through acidified water, it can be split into hydrogen and oxygen.

A compound is different from a mixture. A **mixture** can contain its components in any proportion, whereas in a **compound**, the component elements are always in a fixed proportion. There are also other differences between a compound and a mixture:

1) No chemical action takes place when a mixture is made. When a compound is made, a chemical action takes place.
2) The components of a mixture can be separated easily by physical means, whereas a chemical reaction is required to separate a compound into its components.

Table 2.3 Symbols of elements.

Element	Symbol
Aluminium	Al
Calcium	Ca
Carbon	C
Chlorine	Cl
Chromium	Cr
Copper	Cu
Fluorine	F
Gold	Au
Helium	He
Hydrogen	H
Iodine	I
Iron	Fe
Lead	Pb
Magnesium	Mg
Manganese	Mn
Mercury	Hg
Nickel	Ni
Nitrogen	N
Oxygen	O
Phosphorus	P
Potassium	K
Silicon	Si
Silver	Ag
Sodium	Na
Sulphur	S
Tin	Sn
Zinc	Zn

2.4 Symbols and Formulae

In chemistry, symbols and formulae are used to represent elements and compounds, respectively. For elements, either one or two letters are used, for example, the letter I is used for Iodine, Al for aluminium and Zn for zinc. For some elements, the symbols are chosen from their Latin names, for instance:

Fe from Ferrum (iron)
Na from Natrum (sodium)
Au from Aurum (gold)

The formula of a compound shows the ratio of its component elements. The formula for water is H_2O, which means that for every atom of oxygen, two atoms of hydrogen are required to form one molecule of water. The reason for this is that the valency of hydrogen is 1 (H^+) and that of oxygen is 2 (O^{2-}).

The valency of a copper ion is 2 (Cu^{2+}) and that of a chloride ion is 1 (Cl^-). Therefore, one copper ion combines with two chloride ions to form $CuCl_2$ (copper chloride).

Symbols used for some of the elements have already been given in Table 2.3; the formulae of some of the compounds are given in Table 2.4.

Table 2.4

Compound	Formula
Aluminium oxide	Al_2O_3
Calcium sulphate	$CaSO_4$
Calcium carbonate	$CaCO_3$
Calcium chloride	$CaCl_2$
Magnesium carbonate	$MgCO_3$
Sodium chloride	$NaCl$
Sodium carbonate	Na_2CO_3
Copper chloride	$CuCl_2$
Copper sulphate	$CuSO_4$
Zinc oxide	ZnO
Ferrous sulphate	$FeSO_4$

Figure 2.8

2.5 Acids and Bases

Acids and bases are substances that have corrosive action on skin, metals and other building materials. The labels of bottles/containers containing acids/alkalis include the sign shown in Figure 2.8.

2.5.1 Acids

There are two types of acids: organic acids and mineral acids. An organic acid is obtained from plant and animal sources; for example, citric acid is obtained from oranges and lemons, and ethanoic acid

Table 2.5

Organic acid	Formula	Mineral acid	Formula
Oxalic acid	CH_3COOH	Hydrochloric acid	HCl
Ethanoic acid	CH_3CO_2H	Nitric acid	HNO_3
		Sulphuric acid	H_2SO_4

from vinegar. Organic acids are weak as compared to mineral acids. Mineral acids, which are strong and highly corrosive, include sulphuric acid, hydrochloric acid and nitric acid.

The chemical formulae of some of the acids are given in Table 2.5.

In general, acids have the following properties:

1) They change the colour of substances called indicators. Blue litmus, an indicator, becomes red when added to an acid.
2) They have a sour taste.
3) Hydrogen ions are released when an acid is added to water.
4) There is a reaction between acids and a large number of metals, producing hydrogen gas.
5) They also react with other materials like cement/concrete.

The reactions between acids and some metals/materials are given below:

Zinc + dilute hydrochloric acid → Zinc chloride + hydrogen gas

Zinc + dilute sulphuric acid → Zinc sulphate + hydrogen gas

Zinc + dilute nitric acid → Zinc nitrate + water + dinitrogen oxide gas

Copper + dilute nitric acid → Copper nitrate + water + nitrogen oxide gas

Iron + dilute hydrochloric acid → Ferrous chloride + hydrogen gas

Iron + dilute sulphuric acid → Ferrous sulphate + hydrogen gas

Aluminium + dilute hydrochloric acid → Aluminium chloride + hydrogen gas

Lead + dilute hydrochloric acid → Lead chloride + hydrogen gas

(A layer of lead chloride will protect the metal from further acid attack.)

Calcium carbonate (marble) + dilute sulphuric acid → Calcium sulphate + water + carbon dioxide gas

2.5.2 Bases

A base is a substance that can neutralise an acid to produce a salt and water. Metal oxides, hydroxides and carbonates neutralise acids and are called **bases**. Alkalis are those bases that dissolve in water, for example, sodium hydroxide and calcium hydroxide. Some of the bases and their formulae are given in Table 2.6.

The properties of bases are:

1) Soluble bases can change the colour of red litmus to blue.
2) They neutralise acids, forming a salt and water.
3) Soluble bases react with oils to form soap.
4) Hydroxide ions are released when an alkali is dissolved in water.

Table 2.6

Base	Formula
Aluminium oxide	Al_2O_3
Copper oxide	CuO
Sodium hydroxide	$NaOH$
Calcium hydroxide	$Ca(OH)_2$
Ammonia solution	NH_3
Copper hydroxide	$Cu(OH)_2$

The following reactions show how bases neutralise acids, resulting in the formation of a salt and water:

Sulphuric acid + sodium hydroxide → Sodium sulphate + water

Copper sulphate + sodium hydroxide → Sodium sulphate + copper hydroxide

Bases can also react with some metals:

Aluminium + concentrated sodium hydroxide + water → Sodium aluminate + hydrogen gas

Exercise 2.1

1 Explain the difference between an element and a compound.

2 Write down the chemical symbols of the following elements:
   A Iron
   B Silver
   C Lead
   D Tin
   E Copper.

3 The following salts produce efflorescence in brickwork. What are the chemical formulae?
   A Calcium chloride
   B Calcium sulphate
   C Magnesium chloride.

4 **A** Explain the difference between organic and inorganic acids.
   B State three properties of acids.

5 **A** State three properties of a base.
   B Give the chemical formula of the base used in houses to unblock the drains.
   C What happens when an acid attacks marble?
   D Complete the following reaction:
   Marble + dilute sulphuric acid →........................+ water +....................

6   Complete the following reactions:
    A Sodium + dilute hydrochloric acid → .... + hydrogen gas
    B Sodium + dilute sulphuric acid → .... + hydrogen gas
    C Iron + dilute hydrochloric acid → Ferrous chloride + ...
    D Iron + ... → Ferrous sulphate + hydrogen gas
    E Aluminium + ... → Aluminium chloride + hydrogen gas
    F Sulphuric acid + sodium hydroxide → .... + water

References/Further Reading

1 Fullick, A. and McDuell, B. (2008). *Edexcel AS Chemistry*. Harlow: Pearson Educational.
2 Fullick, P. (2006). *AQA Science – GCSE Chemistry*. Cheltenham: Nelson Thornes.

3

Effects of Chemicals and the Atmosphere on Materials

LEARNING OUTCOMES
1) Explain oxidation and its effects on metals. 2) Describe electrolysis and electrolytic corrosion. 3) Explain the applications of electrolysis.

3.1   Introduction

Metals fulfil an important role in the construction of buildings. Aluminium, steel, lead and copper have a range of uses in the construction of buildings and other structures. Some of the uses are:

Aluminium: doors, windows, cladding and roof construction
Steel: doors, windows, structural steel sections such as universal beams, universal columns etc. and reinforcement for reinforced concrete
Lead: roof construction
Copper: pipes for central heating and water supply, and roofing

When exposed to the atmosphere these metals are affected by air, water and impurities in the air, and the phenomenon is known as **corrosion**. There are two ways in which the corrosion of metals can occur: oxidation and electrolytic corrosion.

3.2   Oxidation

Oxidation is the action of oxygen and water on metals resulting in the formation of a new substance. An example of this effect that we come across quite often is the corrosion of iron. This phenomenon is also known as **rusting** and is due to the oxidation of iron forming a new substance.

Iron + water + oxygen $\rightarrow$ hydrated iron oxide

$$4Fe + 2H_2O + 3O_2 \rightarrow 2Fe_2O_3.2H_2O$$

The rust (hydrated iron oxide) formed is reddish-brown in colour and porous. The rust allows more air and water to pass through and attack the metal underneath. Iron/steel is extracted from iron ores that are oxides of iron. When iron/steel components are exposed to the atmosphere, the metal

Construction Science and Materials, Second Edition. Surinder Singh Virdi.
© 2017 John Wiley & Sons Ltd. Published 2017 by John Wiley & Sons Ltd.
Companion website: www.wiley.com/go/virdiconstructionscience2e

reacts with air and moisture to revert to the parent material, i.e. the iron ore. A thin layer of rust can be useful sometimes; for example, a very small amount of rust on reinforcement steel will improve its bond with fresh concrete.

Both air and water are required for rust to form. In the absence of either or both, rusting of iron will not occur. This can be demonstrated by performing a simple experiment, as explained in Section 3.2.1. Iron/steel is used in manufacturing several building components; therefore, it is necessary to provide a protective coating so that rusting can be checked. A detailed discussion on the protection of steel is given in Section 3.4.2.

3.2.1 Experiment: To Show that Oxygen (Or Air) and Water are Necessary for the Rusting of Iron

Apparatus and materials: glass tubes, tap water, boiled water, iron nails and calcium chloride crystals.

Method: Three or four nails are put into test tubes and exposed to three types of environment (see Figure 3.1):

a) Tap water;
b) Water that has been boiled for a few minutes;
c) Dry air.

On checking the nails after a few days, it will be noticed that:

1) The iron nails in the tap water are rusted as they are exposed to water and air. Ordinary water contains some dissolved oxygen.
2) The iron nails in the water that has been boiled and cooled are not rusted. When ordinary water is boiled for some time, most of the dissolved air is removed. For rusting, both air and water are required.
3) The nails in dry air are not rusted. Calcium chloride crystals absorb moisture from the air, making it dry.

If the nails are partially submerged in water, the intensity of rusting will be far greater, as moisture and air are both easily available.

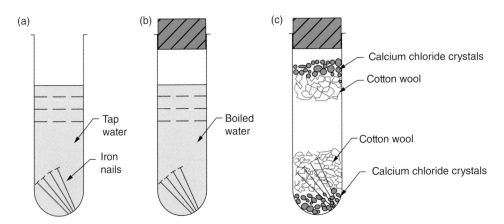

Figure 3.1 Oxidation (rusting) of iron nails in: (a) water and air; (b) boiled water; (c) dry air.

Table 3.1 Effect of the atmosphere on some metals.

Metal	Effect of the atmosphere
Aluminium	Aluminium is covered with a film of aluminium oxide which prevents further corrosion of the metal.
Copper	Copper is covered with a protective green coating which is carbonate/sulphate of copper.
Zinc	A film of basic zinc carbonate is formed which protects the metal from further corrosion.
Lead	The surface of freshly cut lead is bright in appearance, but on exposure to the atmosphere it becomes dull. This is due to the formation of basic lead carbonate which prevents further corrosion.

This experiment proves that for the oxidation of iron, both air and water (moisture) are required. In the absence of either or both, oxidation will not take place.

The corrosive action of the atmosphere is not just limited to steel, other metals are also affected. However, the corrosive action is not drastic as the surface corrosion protects the metals from further deterioration. Table 3.1 summarises the effects of the atmosphere on a selection of metals.

These metals are also affected by acid rain. The effect, however, is insignificant as:

1) The acid is very weak.
2) The protective layer of carbonate/oxide prevents the acid from penetrating and causing a reaction with the metal.

More information on acid rain is given in Section 3.6.

3.3 Electrolysis

It is important to study **electrolysis** before trying to understand electrolytic corrosion, which can deteriorate zinc, iron and some other metals.

In Figure 3.2, the circuit shown consists of:

1) A cell shown as ⊣⊢
2) Metal rods or plates known as **electrodes**. The negative electrode, which is connected to the negative terminal of the cell, is called the **cathode**. The electrode connected to the positive terminal of the cell is called the **anode**. The flow of current from the cell is considered to be from the positive terminal to the negative terminal.
3) A liquid that allows the current to pass through. It is called an **electrolyte**, and is basically a solution in water of acids, alkalis or salts. Non-electrolytes such as pure water do not allow electricity to pass through.

If we take copper sulphate as the electrolyte, then there are two types of ions in the solution: positively charged copper ions and negatively charged sulphate ions. When current is passed through the electrodes, they become charged and, consequently, positively charged copper ions will be attracted towards the cathode which is negatively charged. Similarly, negatively charged sulphate ions are attracted towards the anode (Figure 3.3). The process of decomposition of an electrolyte is known as **electrolysis**.

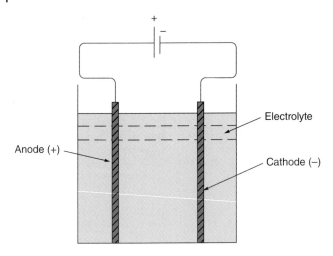

Figure 3.2

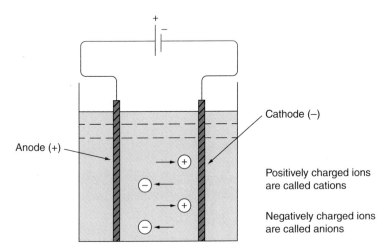

Figure 3.3

Electrolysis can be used in many industrial processes such as electroplating, production of sodium hydroxide, extraction of aluminium from bauxite and purification of copper.

3.4 Electrolytic Corrosion

As discussed in Section 3.3, the passage of electric current through an electrolyte can start a chemical action. Similarly, a chemical action can cause the flow of an electric current when two metals, placed in an electrolyte, are in contact. A situation similar to this may arise in buildings or other structures where two dissimilar metals have been used. In the presence of rainwater, which is slightly acidic and hence an electrolyte, an electric circuit is formed which results in the corrosion of one of the metals.

The rate at which a metal is dissolved (or corroded) depends on the relative position of the metal in the electrochemical series. Table 3.2 shows the electrode potential of a selection of metals.

Metals at the top of the table are more reactive (or easily corroded) whereas silver and gold are highly resistant to corrosion. Each metal when immersed in an electrolyte attains a potential which is shown in Table 3.2. If plates or rods of zinc and copper are immersed in an electrolyte and connected by a wire, the voltage will be approximately 1.1 volts.

$$\text{Voltage} = 0.34 - (-0.76) = 1.1 \text{ volts}$$

The phenomenon of electrolytic corrosion can be understood by studying the simple cell shown in Figure 3.4.

Table 3.2 The electrochemical series.

Metal	Valency	Electrode potential (volts)
Calcium	Ca^{2+}	−2.87
Magnesium	Mg^{2+}	−2.34
Aluminium	Al^{3+}	−1.66
Zinc	Zn^{2+}	−0.76
Iron	Fe^{2+}	−0.44
Tin	Sn^{2+}	−0.14
Lead	Pb^{2+}	−0.13
Copper	Cu^{2+}	+0.34
Silver	Ag^{2+}	+0.80
Gold	Au^{3+}	+1.50

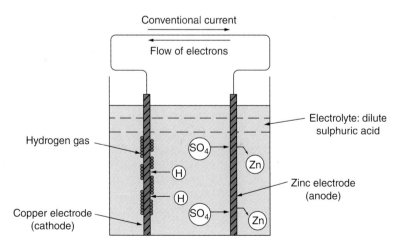

Figure 3.4 The simple cell.

The zinc electrode can be amalgamated with mercury so that the acid does not attack zinc directly, starting a reaction between the two. Zinc is more reactive than copper, therefore some of its atoms pass into solution before those of copper, leaving electrons behind. Each atom leaves two electrons on the zinc electrode:

$$Zn \rightarrow Zn^{2+} + 2e^-$$

Zinc becomes negatively charged because of the electrons left behind.

Sulphuric acid ionises into hydrogen and sulphate ions when it is mixed with water:

$$H_2SO_4 \rightarrow 2H^+ + SO_4^{2-}$$

The sulphate ions are attracted towards zinc ions to form zinc sulphate ($ZnSO_4$). As sulphate ions are attracted by the zinc ions, hydrogen ions move and deposit on the copper electrode. Electrons that have been left on the zinc electrode flow towards the copper electrode, as shown in Figure 3.4. Hydrogen ions, deposited on the cathode, receive these electrons, resulting in the formation of hydrogen gas. As more zinc ions enter the solution from the zinc electrode, corrosion occurs. Corrosion will be severe if the size of the zinc electrode is very small as compared to the copper electrode. The flow of electrons through the wire is called an electric current.

When the cell was first discovered, scientists thought that the current flow was from the positive to the negative electrode. This convention is still used to show the flow of current.

In the simple cell, the anode is the negatively charged electrode, which is different from the statement in Section 3.3 that the anode is positively charged. This confusion can be avoided if we define an anode and a cathode as:

anode: the electrode where conventional current enters the cell.
cathode: the electrode where conventional current leaves the cell.

3.4.1   Examples of Electrolytic Corrosion

If the zinc coating is damaged, a simple cell is formed as the two metals are in contact in the presence of water (Figure 3.5). As zinc is more reactive than steel, it becomes the anode and starts to corrode. The zinc coating is called the sacrificial anode as it corrodes to protect the main metal, i.e. steel.

1) Steel water tanks are provided with a coating of zinc, which protects steel from corrosion.
2) In central heating systems that have steel radiators and copper pipes, steel becomes the anode as it is higher than copper in the electrochemical series. The joint between a radiator and a copper pipe may leak after a number of years. In more severe cases, pin holes may appear in the radiators.

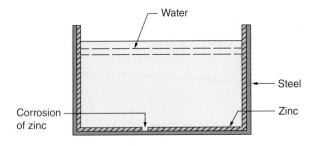

Figure 3.5

3) Mild steel consists of two grains: ferrite and pearlite. Ferrite is pure iron but pearlite is formed from alternate layers of ferrite and cementite, Fe_3C. Ferrite is known to be anodic to pearlite and starts to dissolve preferentially in the presence of water. There are also other reasons for the production of anodes and cathodes in steel, for example, the inclusion of some other grains in steel, surface imperfections etc.

4) Lead is an extremely malleable metal and easily moulded into any shape. It is used as a roofing material to cover large, as well as small, areas. Correct fixing of leadwork is necessary to avoid premature failure. Lead sheets should be fixed by copper nails, as galvanised iron nails will corrode due to electrolytic corrosion.

3.4.2 Protection of Steel from Corrosion

1) **Galvanising:** The steel components or objects are coated with a thin layer of zinc or tin, either by electroplating or by dipping in the molten metal. Zinc and tin are not affected by oxidation. However, the performance of the protective layer is affected if it is scratched and there is water present.

2) **Surface treatments:** Steel components such as window frames can be protected from corrosion by applying paint; other objects can be covered with bitumen or grease depending on where they are situated. The main disadvantage with this method is that the protective layer has to be renewed every few years.

3) **Sacrificial anode:** A metal can be protected by attaching it to another metal that is higher in the galvanic series. As explained earlier, the metal that is higher in the electrochemical series will corrode to protect the other metal. Underground steel pipes are connected to a more electronegative metal such as zinc or magnesium. A simple cell is created which causes the corrosion of zinc/magnesium, protecting the steel pipe. This method is also known as **galvanic cathodic protection**.

4) **Impressed current cathodic protection:** For larger structures such as reinforced concrete, steel piles, gas pipelines etc., the galvanic anodes cannot provide enough current for complete protection. In these cases, impressed current cathodic protection (CP) may be used to protect the structure.

Steel reinforcement is used in concrete to resist tensile stress. The reinforcement bars are protected from corrosion by the alkaline concrete environment. In winter, de-icing salts are used on concrete road surfaces and bridge decks which permeate the concrete gradually. The chloride ions from the de-icing salt neutralise the alkaline concrete environment and thus the reinforcement is exposed to corrosion. The corrosion of steel reinforcement can be prevented by impressed current CP which involves applying an electrical field to the concrete structure with the steel reinforcement being used as the cathode, as shown in Figure 3.6. The current will flow from the anode to the steel reinforcement, protecting the steel from corrosion.

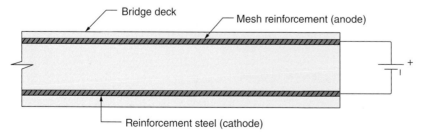

Figure 3.6 Impressed current cathodic protection.

A number of anode systems are available; titanium mesh with concrete overlay is used for horizontal applications such as bridge decks.

3.5 Applications of Electrolysis

Electrolysis is used in several applications, some of which are given here.

3.5.1 Electroplating (Figure 3.7)

Electrolysis can be used to coat iron or other, cheaper metals with thin layers of more expensive metals like silver and gold. If we want to coat an iron nail with copper, it is connected to the negative terminal of the battery, i.e. it is made a cathode. The positive terminal of the battery is connected to a plate or rod of copper. The electrolyte can be a salt of copper, for example, copper sulphate, copper nitrate etc.

When the current is switched on, the copper electrode dissolves to form copper ions:

$$\text{Copper}\,(\text{solid}) \rightarrow \text{Copper}^{2+}\,(\text{aqueous}) + 2\,\text{electrons}^{-}$$

$$\text{or, Cu} \rightarrow \text{Cu}^{2+}\,(\text{aqueous}) + 2e^{-}$$

The electrons flow from the copper electrode to the iron nail. At the cathode, the copper ions in the electrolyte are turned into copper atoms and deposited as a thin layer on the nail.

3.5.2 Extraction of Aluminium

Aluminium is extracted from its ore, bauxite, which is aluminium oxide (Al_2O_3. $2H_2O$). Aluminium oxide is very stable and is not easily reduced to aluminium. Extraction by electrolysis is not easy because the melting point of bauxite is very high (2050 °C). The problem is solved by using molten cryolite (sodium hexafluoroaluminate, Na_3AlF_6) as the electrolyte in which bauxite dissolves at about 900 °C.

Carbon is used as the anode as well as the cathode in the cell. As the current is passed, molten aluminium collects at the bottom of the cell from where it is tapped off, as shown in Figure 3.8.

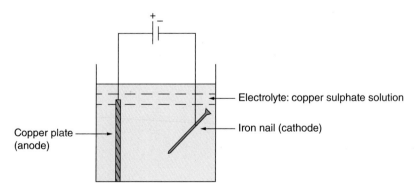

Figure 3.7 **Electroplating.**

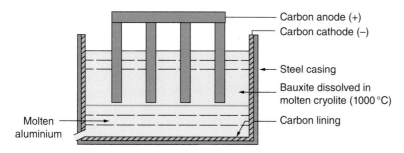

Figure 3.8 Electrolysis of bauxite.

3.6   Acid Rain

Acid rain is a combination of very weak acids such as sulphuric acid and nitric acid. Most of the sulphur dioxide gas that is converted to acid comes from coal when it is burnt in industrial processes such as power generation at thermal power stations. Coal contains up to about 5% sulphur, which is responsible for producing sulphur dioxide gas on burning:

$$\text{Sulphur} + \text{oxygen (in air)} \rightarrow \text{sulphur dioxide gas}$$

In the presence of sunlight, sulphur dioxide, oxygen and water vapour combine to form sulphurous acid. This is further oxidised to form sulphuric acid which falls when it rains or snows.

In a similar way, the exhaust fumes of motor vehicles are converted to nitric acid when they react with water vapour. Rainwater also dissolves carbon dioxide gas (CO_2) from the air to form carbonic acid, H_2CO_3. The pH value of acid rain is about 4.

Acid rain affects trees and buildings built of marble. Dilute acids react with marble to produce hydrogen gas, salt and water. After prolonged attack, acid rain causes discoloration and disintegration of marble, as is happening in the case of the Taj Mahal, the white marble mausoleum built in 1622 in Agra (northern India).

The effects of chemicals and the atmosphere on other materials are discussed in Chapter 16.

References/Further Reading

1 Breithaupt, J. (2008). *AQA Physics A*. Cheltenham: Nelson Thornes.
2 Fullick, P. (2006). *AQA Science – GCSE Chemistry*. Cheltenham: Nelson Thornes.

4

Electricity

LEARNING OUTCOMES

1) Explain Ohm's law and solve numerical problems involving this law.
2) Solve problems involving resistors in series/parallel.
3) Explain the construction of a transformer and solve associated numerical problems.
4) Describe the distribution of electricity from the power plant to the consumer.

4.1 Introduction

When a plastic rod is rubbed with nylon or another synthetic fabric, it becomes electrically charged and will attract light objects like small pieces of paper. Similarly, when a glass rod is rubbed with silk it becomes electrically charged. On rubbing the glass rod with silk, some electrons from the glass are attached to the silk, making the glass positively charged due to the loss of electrons. The charge on the glass rod (and the plastic rod) is called a static electric charge. Any two charged objects will exert a force on each other. Similar charges will produce a repulsive force while opposite charges produce an attractive force.

Electric current is the flow of electric charge through a conductor. Both a static charge and an electric current are formed because of the electrons.

4.2 Coulomb's Law

In 1785, Coulomb, a French physicist, measured the force of attraction and repulsion between two stationary electrical charges and stated that the charges attract or repel each other with a force (F) that is directly proportional to the product of the two charges and inversely proportional to the square of the distance between them:

$$F \propto \frac{q_1 q_2}{r^2}$$

$$\text{or, } F = k\frac{q_1 q_2}{r^2}$$

where q_1 and q_2 are the electric charges and r is the distance between them.
k is the constant of proportionality and is known as the **electrostatic constant**.

Construction Science and Materials, Second Edition. Surinder Singh Virdi.
© 2017 John Wiley & Sons Ltd. Published 2017 by John Wiley & Sons Ltd.
Companion website: www.wiley.com/go/virdiconstructionscience2e

The value of k is $9 \times 10^9 \dfrac{\text{Nm}^2}{\text{C}^2}$

Therefore, $F = 9 \times 10^9 \dfrac{q_1 q_2}{r^2}$

If two charges placed at a distance of 1 m are similar and equal, and repelled with a force of 9×10^9 N, the magnitude of each charge is 1 coulomb.

4.3 Electric Current

An electric current is defined as the rate of flow of electric charge through a circuit. The unit of current in SI units is the ampere (A).

The **ampere** is defined as the current if 1 coulomb of charge flows through a conductor in one second.

$$1 \text{ ampere} \left(A \right) = 1 \text{ coulomb per second}$$

An ammeter is used to measure the electric current through a conductor and must be connected in series with the conductor, as shown in Figure 4.1.

4.4 Potential Difference

To understand potential difference (p.d.), consider a pipe that allows the flow of a liquid. The flow is only possible if the pipe is laid at a slope, i.e. the ends are at different heights. Similarly, current will flow through a conductor if its ends are maintained at different charges, or at different potentials.

The unit of potential difference is the **volt**.

A voltmeter is used to measure the potential difference between two points on a circuit. It is always connected in parallel, as shown in Figure 4.1.

Figure 4.1

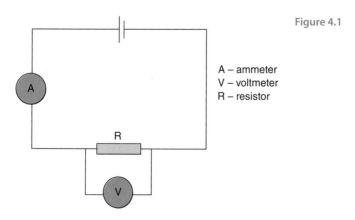

A – ammeter
V – voltmeter
R – resistor

4.5   Electromotive Force (e.m.f.)

The maximum potential difference that a cell can produce in a circuit is called its **electromotive force** or e.m.f. This includes the potential difference required to drive the current through the cell itself. The e.m.f. of a cell depends only on the materials used in its manufacture. As mentioned in Section 4.4, electric current flows in a circuit/conductor if its ends are at different potentials. The e.m.f. may be considered to be the external force necessary to make the current flow in a circuit. It is defined as the work done (in joules) in moving a unit charge from one end of the circuit to the other.

4.6   Ohm's Law

Ohm's law is the most fundamental law of electricity and was formulated by Georg Ohm in 1826. He conducted experiments using several types of wires and found that the current (I) flowing through each depended on the potential difference (V) applied across their ends. The law states that the current flowing through a metal conductor is directly proportional to the p.d. applied across its ends provided that the temperature and other physical conditions remain constant.

$$V \propto I$$
$$V = I.R$$

R is the constant of proportionality known as the **resistance** of the conductor. It is the resistance offered by the conductor to the flow of charge. Its value depends on the material, dimensions and the temperature of the conductor.

The above formula can also be written as:

$$R = \frac{V}{I} \text{ and } I = \frac{V}{R}$$

Resistance is measured in ohms (Ω).

The resistance of a conductor is one **ohm** if a current of one ampere flows through it on applying a p.d. of one volt.

Devices specially made to provide resistance in electrical circuits are known as **resistors**. They reduce the flow of current in a simple circuit. Figure 4.2 shows a variable resistor known as a rheostat which is used for varying the current flowing in a circuit. Moving the position of the sliding contact changes the length of the coiled wire and hence its resistance.

Figure 4.2 Rheostat.

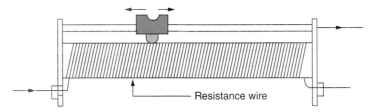

Resistance wire

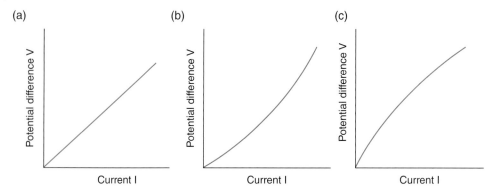

Figure 4.3

The resistance of most metals increases with temperature. However, some materials like silicon become better conductors (resistance decreases) as they get warmer. Figure 4.3 shows three relationships between potential difference and current for metals/materials used in electrical circuits:

- Figure 4.3(a) shows that the current in a metal wire is proportional to the p.d. at a constant temperature.
- Figure 4.3(b) shows that, as the temperature rises, the current does not remain proportional to the p.d. In most metals, an increase in p.d. causes a smaller increase in the current, or, in other words, the resistance increases.
- Figure 4.3(c) shows that in silicon, an increase in the p.d. causes a greater increase in the current, hence the resistance decreases.

Example 4.1 Calculate:

a) The current I in the circuit of Figure 4.4(a)
b) The voltmeter reading in Figure 4.4(b)
c) The value of the resistor in the circuit of Figure 4.4(c)

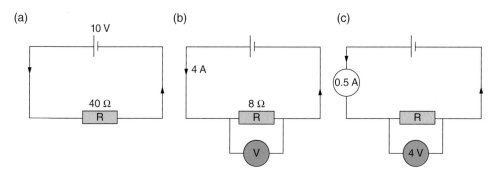

Figure 4.4

Solution:

a) $I = \dfrac{V}{R} = \dfrac{10}{40} = \mathbf{0.25\,A}$

b) $V = I\ R = 4 \times 8 = \mathbf{32\,V}$

This is equal to the voltage read by the voltmeter.

c) $R = \dfrac{V}{I} = \dfrac{4}{0.5} = \mathbf{8\,\Omega}$

4.7 Electrical Resistivity and Conductivity

The resistance of a conductor is directly proportional to its length (L) and inversely proportional to its cross-sectional area (A).

$$\text{Resistance}\ R \propto \frac{L}{A}$$

$$R = \rho\frac{L}{A}\ \text{or,}\ \rho = \frac{AR}{L}$$

ρ, the constant of proportionality, is known as the **electrical resistivity**.

Metals and their alloys have low resistivity and are known as good conductors of electricity. Wood, rubber, glass etc. have high resistivity and are called bad conductors of electricity, or good insulators. Materials with resistivity in between those of conductors and insulators are called semi-conductors. Table 4.1 gives the values of electrical resistivity for a selection of materials.

4.8 Resistors in Series/Parallel

4.8.1 Resistors in Series

Figure 4.5 shows three resistors, R_1, R_2 and R_3, connected in series. If a battery is connected across the circuit, the same current, I, passes through each resistor.

V_1, V_2 and V_3 are the potential differences across resistors R_1, R_2 and R_3, respectively.

$$V_1 = IR_1\ \ V_2 = IR_2\ \ V_3 = IR_3$$

Table 4.1 Electrical resistivity for some metals/materials.

Material	Electrical resistivity at 0 °C Unit: Ωm
Carbon	3.5×10^{-5}
Silver	1.6×10^{-8}
Copper	1.7×10^{-8}
Aluminium	2.7×10^{-8}
Iron	10.0×10^{-8}
Wood	$10^{8}\text{–}10^{11}$
Glass	$10^{10}\text{–}10^{14}$

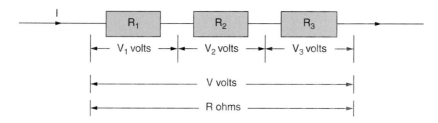

Figure 4.5 Resistors in series.

In a series circuit, the sum of individual voltages across individual resistors is equal to the total applied voltage:

$$V = V_1 + V_2 + V_3$$
$$= I R_1 + I R_2 + I R_3 \tag{4.1}$$
$$= I(R_1 + R_2 + R_3)$$

If R is the total resistance, then:

$$V = I R \tag{4.2}$$

From equations (4.1) and (4.2), $R = R_1 + R_2 + R_3$
The following conclusions can be drawn for resistors in series:

1) The total resistance of a circuit is equal to the sum of the individual resistances and the internal resistance of the battery.
2) The current in all resistors is the same.
3) The p.d. across a resistor depends on its resistance.

4.8.2 Resistors in Parallel

Resistors placed side by side in a circuit are said to be in parallel (Figure 4.6). The potential difference across each resistor is equal to the applied potential difference. If I_1, I_2 and I_3 are the values of current through resistors R_1, R_2 and R_3 respectively, the total current, I, is given by:

$$I = I_1 + I_2 + I_3$$

The potential difference across each resistor is the same, i.e. V. Therefore:

$$V = I_1 R_1 = I_2 R_2 = I_3 R_3$$

$$\text{or, } I_1 = \frac{V}{R_1}; \quad I_2 = \frac{V}{R_2}; \quad I_3 = \frac{V}{R_3}$$

$$I = \frac{V}{R_1} + \frac{V}{R_2} + \frac{V}{R_3}$$

$$= V\left(\frac{1}{R_1} + \frac{1}{R_2} + \frac{1}{R_3}\right)$$

$$\frac{I}{V} = \frac{1}{R_1} + \frac{1}{R_2} + \frac{1}{R_3}$$

$$\frac{1}{R} = \frac{1}{R_1} + \frac{1}{R_2} + \frac{1}{R_3} \quad \left(\frac{I}{V} = \frac{1}{R}\right)$$

where R is the total resistance of the circuit.

Figure 4.6 Resistors in parallel.

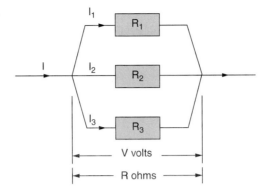

The following conclusions can be drawn for resistors in parallel:

1) The reciprocal of the total resistance of a circuit is equal to the sum of the reciprocals of the individual resistances.
2) The p.d. is the same in all resistors.
3) The current in each resistor depends on its resistance.

Example 4.2   A current of 2 A flows through two resistances of 6 Ω and 4 Ω connected in series. Find the p.d. across each resistance and the supply voltage.

Solution:
Figure 4.7 shows the circuit in which a current of 2 A flows.

Figure 4.7

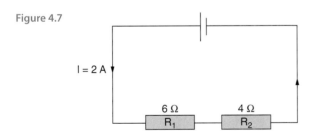

$R_1 = 6\Omega$, and $R_2 = 4\Omega$

Total resistance $R = R_1 + R_2 = 6 + 4 = 10\Omega$

Potential difference across 6Ω resistor, $V_1 = IR_1$
$$= 2.0 \times 6 = \textbf{12.0 V}$$

Potential difference across 4Ω resistor, $V_2 = IR_2$
$$= 2.0 \times 4 = \textbf{8.0 V}$$

Supply voltage, $V = IR$
$$= 2.0 \times 10 = \textbf{20.0 V}$$

Example 4.3

Figure 4.8 shows two resistors connected in parallel in a circuit. Find:

a) The total resistance of the circuit.
b) The current in the circuit.
c) The current in the 8 Ω resistor.

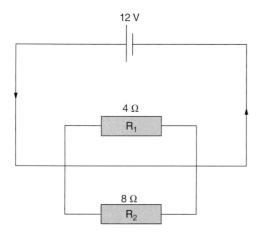

12 V   Figure 4.8

4 Ω
R₁

8 Ω
R₂

Solution:

a) Total resistance, R, can be represented as:

$$\frac{1}{R} = \frac{1}{R_1} + \frac{1}{R_2}$$

$$= \frac{1}{4} + \frac{1}{8}$$

$$\frac{1}{R} = \frac{3}{8}$$

After transposition, $3R = 1 \times 8$ or, $R = \frac{8}{3} = 2.67\,\Omega$

b) Current in the circuit,

$$I = \frac{V}{R}$$

$$= \frac{12}{2.67} = 4.49\ \text{A}$$

c) Current in the 8 Ω resistor $= \dfrac{V}{R}$

$$= \frac{12}{8} = 1.5\ \text{A}$$

4.9 Transformers

Transformers are used to increase or decrease the voltage so that some electrical devices can work using the mains supply of 230 V. For example, the voltage is reduced from 230 V for use with calculators, laptop computers, power tools and other devices. There are two types of transformers:

- Step-up transformers: to increase the voltage;
- Step-down transformers: to decrease the voltage.

A transformer consists of two coils (or windings) of copper wire, the primary and the secondary coil, wound around a soft iron core (Figure 4.9). The primary coil is supplied with alternating voltage from a.c. supply, which creates a magnetic field around the coil. The magnetic field is continuously changing due to the nature of the a.c. supply. The changing magnetic field passes through the secondary coil and induces alternating voltage in it. This process, which is an example of electromagnetic induction without using a permanent magnet, is known as **mutual induction**. Only varying voltage can cause mutual induction, which means that only a.c. supply can be used as the primary input.

Figure 4.10(a) shows that in a step-up transformer, the number of turns in the primary coil is fewer than the number of turns in the secondary coil. In a step-down transformer (Figure 4.10(b)), the arrangement of coils is reversed. The voltage induced in the secondary coil depends on:

1) The number of turns in the primary coil;
2) The number of turns in the secondary coil;
3) The voltage in the primary coil.

Figure 4.9 Transformer.

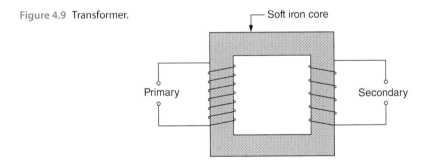

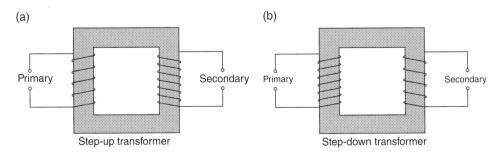

Figure 4.10 (a) Step-up and (b) step-down transformers.

These factors can be combined to produce a formula:

$$\frac{\text{Primary voltage}\,(V_p)}{\text{Secondary voltage}\,(V_s)} = \frac{\text{Number of turns in the primary coil}\,(N_p)}{\text{Number of turns in the secondary coil}\,(N_s)}$$

or, $\dfrac{V_p}{V_s} = \dfrac{N_p}{N_s}$

The materials used for the windings (copper) and the core (soft iron) are selected so that the resistance offered by the windings to the current is low to avoid heating of the coil and thus wastage of energy. Soft iron is magnetised easily on passing the current, and, similarly, is demagnetised quickly when the current flow is stopped. The core is constructed of laminated layers of insulated soft iron to stop the flow of current between the layers and thus keep the losses due to the heating effect to a minimum. An efficiently designed core ensures that all the electrical energy given to the primary coil is passed on to the secondary coil.

Power input in primary = Power output from secondary

$$V_p \times I_p = V_s \times I_s \left(\text{power} = V \times I\right)$$

where I_p and I_s are the currents in the primary and the secondary coil, respectively.

Transformers are used in the transmission of electrical energy from a power station to the consumer. They are initially used to step up the voltage, and then to step it down gradually.

Example 4.4   A transformer steps down the main supply of 230 V to 9 V to operate a calculator.

a) If the secondary coil has 80 turns, calculate how many turns are on the primary coil.
b) Calculate the current in the primary if the current in the calculator is 2 A. Assume that the transformer is 100% efficient.

Solution:

a) $\dfrac{V_p}{V_s} = \dfrac{N_p}{N_s}$

$\dfrac{230}{9} = \dfrac{N_p}{80}$

$N_p \left(\text{number of primary turns}\right) = \dfrac{230 \times 80}{9} = \mathbf{2044.4\ or\ 2045}$

b) $V_p \times I_p = V_s \times I_s$

$230 \times I_p = 9 \times 2$

$I_p \left(\text{current in the primary coil}\right) = \dfrac{9 \times 2}{230} = \mathbf{0.078\ A}$

4.10   Power Generation

Electrical energy is produced from several resources, which can be grouped into two broad types: renewable and non-renewable, as shown in Table 4.2.

Table 4.2

Renewable resources	Non-renewable resources
Geothermal	Coal
Hydroelectricity	Natural gas
Solar	Nuclear fuels (uranium and plutonium)
Tides	Oil
Waves	
Wind	

There are advantages and disadvantages of both types of resources; however, the disadvantages of non-renewable resources are more serious. The fossil fuels (coal, oil and natural gas) will eventually run out as their deposits are limited. To stop the fossil fuels running out so quickly, we need to conserve energy and use more renewable resources. Whilst most resources require turbines and generators for producing electricity, a wind turbine has its own generator.

The majority of power stations use a source of energy (coal, water etc.), turbines and generators. In the case of fossil fuels and nuclear energy, steam is produced in boilers, which is used to drive the turbines. In **hydroelectric** power stations, water is stored by building a concrete, or an earth dam. Water is allowed to fall through large-diameter pipes onto the turbines. The impact of the water has enough energy to drive the turbines. A similar arrangement is used to generate electricity from tidal barrages.

Wind turbines built in exposed places are used to produce electricity directly. The wind turns the blades of the turbine, which turns the generator that is provided inside each turbine.

There are three ways in which **solar energy** can be harnessed: solar cells (or photo-voltaic cells); solar panels; and solar furnaces. Solar cells generate electric current directly from sunlight. Solar panels contain water pipes under a black surface. Heat is absorbed from the sun by the black surface, which heats the water in the pipes. A solar furnace consists of a large array of curved mirrors, which focus the sunlight onto one spot. This produces a very high temperature, which is used to produce steam from water. Steam is used to drive turbines as in the other power stations.

The governments and people of most nations now realise that we need renewable resources of energy as the reserves of coal and crude oil may only last for a few decades more. The governments of some countries are very serious about the energy issue and are investing large amounts of money in solar power and other renewable resources. Abu Dhabi (United Arab Emirates) has developed a new town, called Masdar City, where electricity is produced from solar power rather than from fossil fuels. Algeria has already developed a 150 MW CSP (concentrated solar power) plant with plans for exporting up to 6 GW of solar power to southern Europe by 2020.

4.11   Power Distribution

In Britain, electricity is generated at a voltage of 11–25 kV and then stepped up to 275 kV or 400 kV by transformers. The power is then transmitted all over the country through the national grid. The UK's **national grid** is a network of pylons and cables and is used to take electrical energy from power

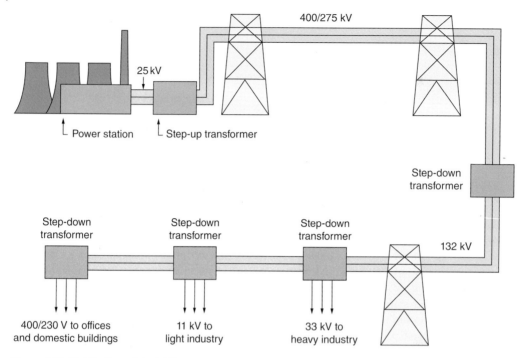

400/275 kV

25 kV

Power station   Step-up transformer

Step-down
transformer

Step-down
transformer

Step-down
transformer

Step-down
transformer

132 kV

400/230 V to offices
and domestic buildings

11 kV to
light industry

33 kV to
heavy industry

Figure 4.11 Distribution of electricity.

stations and supply it to homes, factories, offices and other buildings/structures. Figure 4.11 shows the distribution of electricity from a power plant to homes, offices and factories. The voltage is stepped down to 11 kV in stages at sub-stations for large industrial sites and to 400 V for general distribution. Connection between any two phase wires provides a 400 V supply, suitable for heavy machinery. Connection between a phase wire and the neutral produces a potential of 230 V. Dwelling houses receive a single-phase supply from one phase wire and the neutral. Buildings receiving a 400 V supply have all three phase wires.

Electricity is generated from coal, water and other resources at a power station. As electricity is distributed over large parts of the country by high-tension overhead power lines, it is important to reduce the energy loss due to heating effects. This can be achieved by supplying electricity at a high voltage, as shown in Example 4.5.

The voltage produced in alternating current (AC) is constantly changing, as shown in Figure 4.12. The maximum, average and RMS (root mean square) values are different from one another, as shown by the relationships:

$$\text{Average voltage} = 0.637 \times \text{maximum voltage}$$

$$= 0.637 \times 325.3 = 207.2 \text{ V}$$

$$\text{RMS voltage} = 0.7071 \times 325.3 = \textbf{230 V}$$

Figure 4.12 Alternating current.

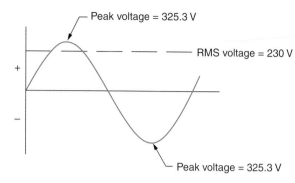

Example 4.5 Find the power wasted as heat in cables when 15 kW is transmitted through a cable of resistance 0.5 Ω at:

a) 230 V.
b) 275000 V.

Solution:

a) $\text{Current} = \dfrac{\text{Power}}{\text{Voltage}} = \dfrac{15000}{230}$

$= 65.2 \, \text{A}$

The power lost in the cable $= I^2 R$

$= 65.2^2 \times 0.5 = \textbf{2125.52 W}$

b) $\text{Current} = \dfrac{15000}{275000} = 0.05454 \, \text{A}$

The power lost in the cable $= I^2 R$

$= 0.05454^2 \times 0.5 = \textbf{0.0015 W}$

The above calculations show that at a voltage of 275000 V the power loss is negligible.

4.12 Supply to Small Buildings

An underground electricity supply cable containing one phase wire and one neutral wire is supplied by the electricity board. The service cable is terminated at a sealed chamber which contains a 100 amp fuse and a neutral link. A connection is made to the meter from which cables are connected to the consumer's unit. Each circuit is protected by a miniature circuit breaker, also called a **fuse**, as shown in Figure 4.13. The lighting circuit and the ring circuit are also shown in Figure 4.13. The ring circuit is used to supply electricity to the power sockets. The lighting circuit is protected by a 6 amp fuse whereas the ring circuit is protected by a 32 amp fuse. There are also fuses for an immersion heater, electric shower and a cooker.

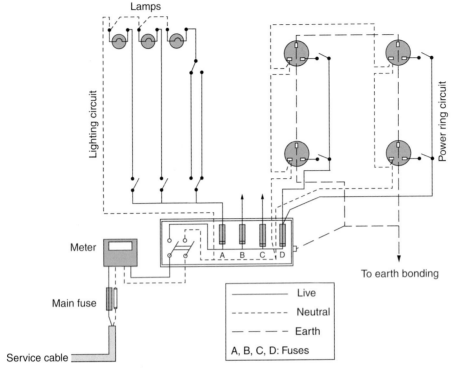

Figure 4.13 Domestic electric installation.

Exercise 4.1

1   A current of 10 A flows through a resistance of 40 Ω. What is the p.d. across this resistance?

2   Two resistances of 4 Ω and 8 Ω are connected in series. Find the p.d. across each resistance and the supply voltage, if a current of 4 A flows through the circuit.

3   Three resistances of 4 Ω, 6 Ω and 12 Ω are connected in parallel. Find the total resistance of the circuit.

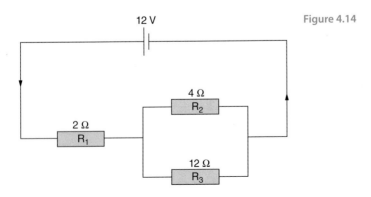

Figure 4.14

4   Two resistances of 4 Ω and 12 Ω are connected in parallel across a 12 V supply. Calculate:
    A The total resistance of the circuit.
    B The current in the circuit.
    C The current through each resistance.

5   Figure 4.14 shows a 12 V battery of internal resistance 1.0 Ω connected to resistors R_1, R_2 and R_3. Find:
    A The total resistance of the circuit.
    B The current in each resistor.

6   A transformer steps down the main supply of 230 V to 110 V to operate a hammer drill.
    A If the primary coil has 115 turns, calculate how many turns are on the secondary coil.
    B Calculate the current in the primary if the current in the hammer drill is 10 A. Assume that the transformer is 100% efficient.

Reference/Further Reading

1   Breithaupt, J. (2008). *AQA Physics A*. Cheltenham: Nelson Thornes.

5

Introduction to Construction Technology

LEARNING OUTCOMES

1) Explain the difference between substructure and superstructure.
2) Discuss a range of foundations for low-rise buildings.
3) Explain the functions of the external envelope and describe the main building elements.

5.1   Introduction

In this chapter, the basic terminology used in building construction and some of the main elements of buildings will be explained. These terms and simple construction processes will be used in other chapters while discussing the effects of heat, sound, forces and light.

Buildings may be constructed in several forms and it is not possible to explain every type of construction here. However, the main forms of construction, such as cellular construction, framed construction and portal frames, will be discussed in this chapter.

5.2   Substructure and Superstructure

The majority of buildings consist of two main parts: the substructure and the superstructure. The **substructure** is usually considered the part which is below ground level (G.L.) and consists of foundations, walls and floors that provide support for the superstructure. The **superstructure** is defined as the part of the building that is above the substructure.

It is necessary to excavate trenches until stable soil strata are reached and the foundation of a building is constructed. In its simplest form, the foundation is a strip of concrete (minimum thickness = 150 mm) that supports the walls of a building. From the foundation, the loading is distributed onto the soil strata, also known as the natural foundation. There are many types of subsoil, each having some ability to carry the building load. This ability is known as the **bearing capacity** of the soil and is defined as the safe load that a unit area of ground will carry. The amount of force exerted by a building on the ground per unit area is known as the **bearing pressure**, and for the stability of the building it should not exceed the bearing capacity. Table 5.1 gives the values of bearing capacity for some soils/rocks.

Construction Science and Materials, Second Edition. Surinder Singh Virdi.
© 2017 John Wiley & Sons Ltd. Published 2017 by John Wiley & Sons Ltd.
Companion website: www.wiley.com/go/virdiconstructionscience2e

Table 5.1 Typical values of bearing capacity.

Subsoil type	Bearing capacity (kN/m^2)
Soft clay	80
Firm clay	120
Fine sand, loose and dry	100
Hard clay	400
Compact gravel and sand	500
Rocks	600–10000

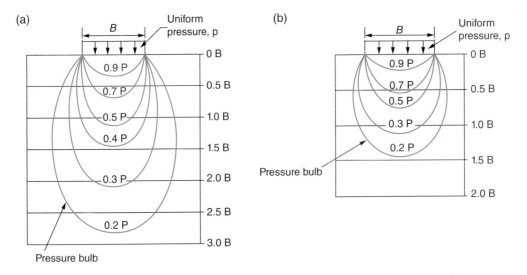

Figure 5.1 Vertical stress under foundations: (a) under a strip foundation; (b) under a pad foundation.

5.2.1 Soil Investigation

Before a building is built, it is important to carry out a soil investigation to find the nature and thickness of the subsoil strata. The results of the subsoil investigation are used for many purposes, one of which is to design the building foundation in a safe and economical manner. During soil investigation, disturbed and undisturbed soil samples are taken to determine the properties of the subsoil and the thickness of the soil strata. Undisturbed samples of soils, taken from boreholes and trial pits, are also used to determine the strength of the soils that will support the structure.

The depth to which the soil investigation should be carried out depends on the type of foundation. Boussinesq's equation is used to determine vertical stresses under continuous, rectangular and circular foundations. Figure 5.1 shows the magnitude of vertical pressure in terms of the bearing pressure, p, at various points. For example, the vertical pressure at any point along the '0.3p line' is equal to 30 per cent of the applied contact pressure. These lines of equal pressure are bulb-shaped and hence are called **pressure bulbs**. The most commonly used pressure bulb for deciding the maximal depth of soil investigation is the one for 0.2p, because practically any stress less than 0.2p is of little consequence. For a typical strip foundation (width (B) = 0.6 m), to be provided at a depth of 1.0 m below ground level, the soil investigation should be done to a depth of 1.0 + 2.8 B, or 2.7 m below G.L.

5.3 Foundations

The function of a foundation is to transmit the total load from a building to the ground safely and without causing any settlement or movement of any part of the building. The foundation of a building should also ensure the safety of the adjoining structures. The foundations of domestic buildings are constructed of concrete because fresh concrete is semi-liquid and will take the shape of the excavation. The most common types of foundations for low-rise buildings (height up to 9.0 m) are strip, deep-strip or trench-fill, wide-strip, raft and pad foundations.

- **Strip foundation (Figure 5.2):** A strip foundation consists of a strip of concrete with uniform width and thickness supporting the full length of all load-bearing walls. Approved document A to the Building Regulations gives suitable minimum widths of strip foundations for various loadings

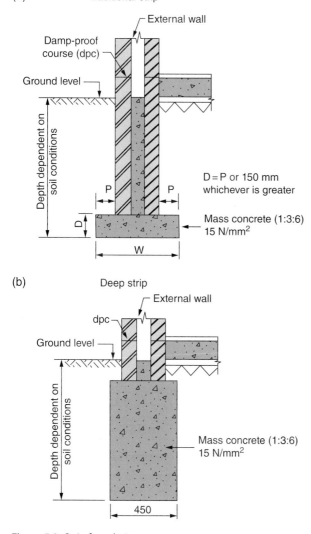

Figure 5.2 Strip foundations.

and types of subsoil. The thickness of concrete should be equal to its projection from the wall, but in no case less than 150 mm. A variation of a strip foundation is that of the deep-strip or trench-fill foundation. More concrete is used in a deep-strip foundation but the amount of brickwork and labour are greatly reduced.

- **Raft foundation (Figure 5.3):** A raft foundation may be used when the bearing capacity of the soil is very low. It covers the whole area of the building and consists of a reinforced concrete slab with extra thickening under load-bearing walls.
- **Pad foundation (Figure 5.4):** A pad foundation is basically a slab of concrete that supports a column. As the columns carry heavy loading, pad foundations are invariably reinforced with steel bars. The size of a pad foundation depends on the column load and the bearing capacity of the subsoil on which it rests.

5.3.1 Settlement

The **settlement** of a structure, which may be caused by several factors, is a result of the compression of the soil strata supporting that structure. Due to the construction of a new building, the stress in the soil supporting the foundation increases. This, in turn, causes the soil particles to come closer,

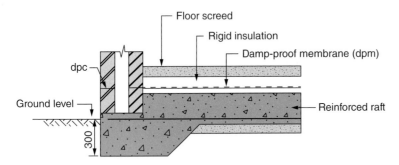

Figure 5.3 Reinforced concrete raft foundation.

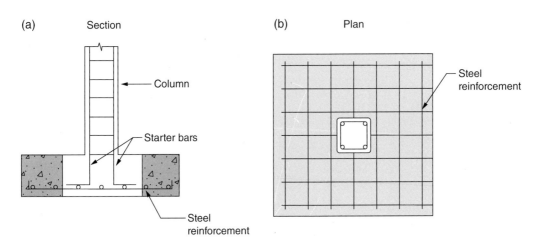

Figure 5.4 Pad foundation.

increasing the contact area between them. The increased inter-particle contact area is responsible for supporting the increased load due to the new construction. As the soil particles come together, the water that is present in the voids, called pore water, is drained out. This process is called the **consolidation of soils**. In sandy soils, the drainage of the pore water is fast – hence the settlement is over in a short period of time. In clay soils, the drainage of the pore water takes a long time – hence the consolidation process is usually slow.

5.4   Forms of Construction

Buildings are provided so that we can perform a range of activities. Houses, shops, schools/colleges, cinema halls, temples, churches, airports etc. are built to cater for the needs of humans. Various forms of construction are available to construct these and other types of buildings; some of these forms of construction are:

- **Cellular construction (Figure 5.5)**: This form of construction consists of load-bearing external and internal walls. The walls spread the loading on strip, or other suitable, foundations which, in turn, disperse it to the subsoils. Buildings constructed using this form are rigid and stable.
- **Cross-wall construction (Figure 5.6)**: This form of construction consists of a series of parallel walls, built at right angles to the front elevation, which carry the floor and roof loads. The floors are built rigidly into the walls to make the walls stable by providing lateral restraint.
- **Framed or skeletal construction**: A series of vertical and horizontal structural members set at right angles, as shown in Figure 5.7, provide support to the roof, floors and walls. The beams support the floors and transfer their load to the columns. The columns transfer the entire load of the building to the foundations. Most multi-storey buildings are constructed using either reinforced concrete (R.C.) or structural steel frames.
- **Portal frames**: Portal frames consist of two uprights rigidly jointed by a horizontal, sloping or curved third member. Each frame requires lateral support, which is provided by bracing from other similar frames, as shown in Figure 5.8.

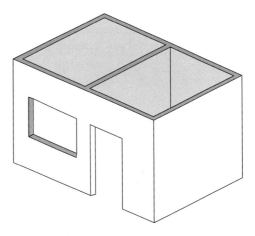

Figure 5.5   Cellular construction.

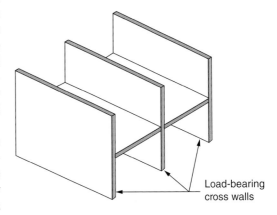

Load-bearing cross walls

Figure 5.6   Cross-wall construction.

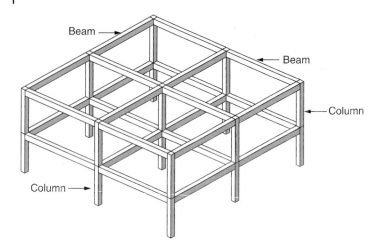

Beam

Beam

Column

Column

Figure 5.7 Framed construction.

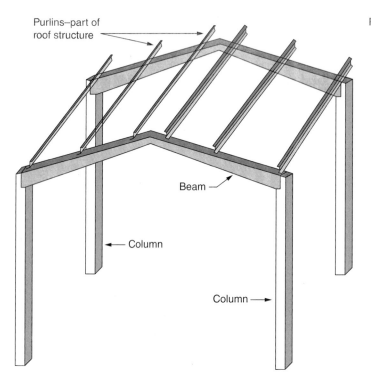

Purlins–part of
roof structure

Beam

Column

Column

Figure 5.8 Portal frames.

5.5   The External Envelope

All human beings require an enclosed space to provide shelter from the elements, and to provide a suitable internal environment for working, rest, recreation and other activities. The shelter can be called the **external envelope** and consists of a roof, walls, doors and windows. A brief description of the main elements of a building is given in Sections 5.5.2 to 5.5.5.

5.5.1 Functions of the External Envelope

There are a number of functions that the external envelope must fulfil; these depend on the type of the structure, the geography of the area and the use for which the building is intended. Some of the main functions are:

- Strength and stability;
- Durability;
- Weather exclusion;
- Fire resistance;
- Thermal insulation;
- Sound insulation.

Strength and stability

The walls and roof of a building must be strong enough to carry the loads (or forces) which may be imposed on them, without excessive deformation. The following contribute towards the total loading:

- The self-weight of the envelope;
- Loads from the internal floors and furniture;
- Loads from the inhabitants;
- Externally applied loads such as wind, rain and snow.

Figure 5.9 shows how the forces are transferred from the roof, walls, floors etc. to the building foundations, and then from the foundations onto the subsoil.

Figure 5.9 Transfer of load from superstructure to substructure.

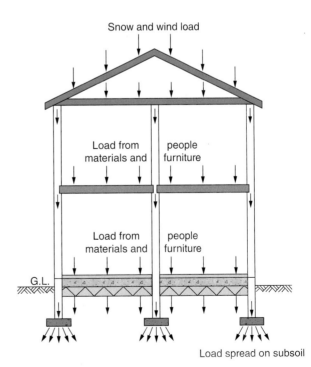

Durability

Building design and construction take a lot of time and money; therefore, the building should require minimal maintenance for its expected useful life. The materials from which the external envelope is constructed must have sufficient resistance to the damaging effects of the loads, solar radiation, wind, frost, rain and chemicals in the environment.

Weather Exclusion

The internal environment of a building must be comfortable for the inhabitants, and to provide this, the materials used for constructing the external envelope must resist the damaging effects of wind and rain.

Fire Resistance

Fires cause loss of human life and damage to buildings. The most important consideration is the safety of human life in the event of an outbreak of fire; therefore, the external envelope must be strong enough to withstand the damaging effects of fire for a period of time to allow the occupants to evacuate the building. Also, the external envelope should contain the fire and prevent it from spreading to other buildings for as long as possible.

Thermal Insulation

For about eight months of the year the external air temperature (in the UK) is lower than the design internal air temperature of a building. This causes a loss of heat energy as heat will flow from the inside of a building to the outside; therefore, it is very important that the external envelope has adequate thermal insulation to keep the heat loss to an absolute minimum.

Sound Insulation

Any unwanted sound is known as noise. Depending on the intensity of noise, it can cause irritation, loss of concentration and health problems. It is not possible to eliminate noise completely, but the choice of materials for constructing the external envelope should be such that the noise from external sources is reduced to an acceptable level.

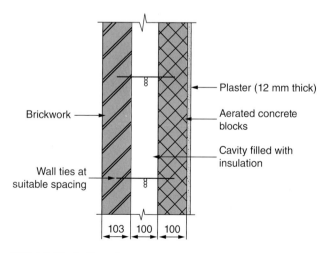

Figure 5.10 Cavity wall.

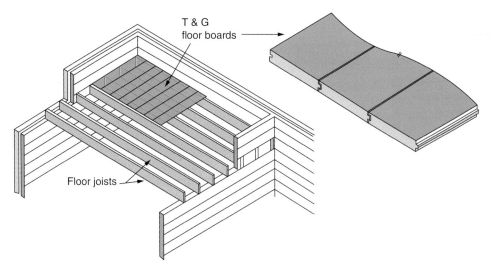

Figure 5.11 Suspended timber upper floor.

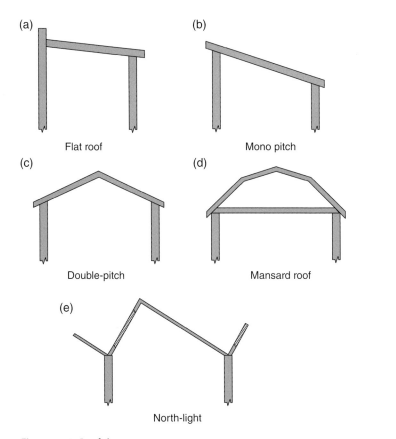

Figure 5.12 Roof shapes.

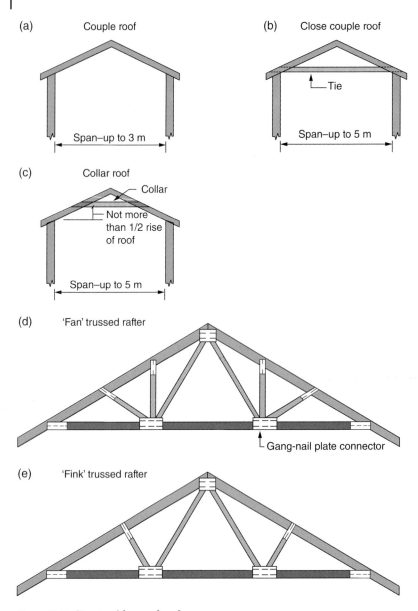

(a) Couple roof
Span–up to 3 m

(b) Close couple roof
Tie
Span–up to 5 m

(c) Collar roof
Collar
Not more than 1/2 rise of roof
Span–up to 5 m

(d) 'Fan' trussed rafter
Gang-nail plate connector

(e) 'Fink' trussed rafter

Figure 5.13 Structural forms of roofs.

5.5.2 Ground Floors

A ground floor may be constructed as either a solid floor or a suspended floor. A solid ground floor (see Figure 5.3) rests on the ground whereas a suspended floor rests on walls. A suspended timber ground floor is not much different from upper-floor construction except that insulation is necessary to have a low U-value.

5.5.3   Cavity Walls

Cavity walls provide better thermal insulation and resistance to moisture penetration as compared to solid brick walls. A cavity wall is constructed by building two walls, or leaves, tied by galvanised steel or plastic wall ties (Figure 5.10). The cavity between the outer brick leaf and inner aerated concrete block leaf is filled with insulation material to conform to the current Building Regulations on thermal insulation.

5.5.4   Suspended Timber Upper Floors

The upper-floor construction consists of timber joists and floor boards or chipboard flooring, as shown in Figure 5.11. The width of the joists is usually 50 mm, but their depth depends on the room span. The thickness of the floor boards or chipboard flooring depends on the spacing between the joists. The joists are supported either directly on the walls or on joist hangers. Plasterboard is fixed to the underside of the joists to provide the ceiling finish for the lower floor.

5.5.5   Roofs

The main function of a building roof is to keep out the rain, snow, wind and sun, and to drain away rainwater. Roofs may be classified by considering their shape, span and structural design. Two of these are explained below:

- **Shape (Figure 5.12):** Some of the main roof shapes are: flat, pitched (mono-pitch, double-pitch, mansard, north-light etc.) and curved (barrel-vault, dome, cross-vault, hyperbolic-paraboloid etc.).
- **Structural form (Figure 5.13):** This method of classification depends on how the roof coverings are supported. Some roof forms considering the structural aspect are: couple roof, close-couple roof, collar roof, purlin roof, trussed-rafter roof, portal frames etc.

When the span of a roof exceeds 5.0 m, the rafters may sag under the weight of the roof coverings. Longitudinal beams called **purlins** are provided to support the rafters, and hence the roof construction is known as a **purlin roof**.

Trussed rafters, used in modern houses, are made by joining the timber sections with gang-nail plates. Trussed rafters are made in factories and are much easier to make as compared to the traditional roof. Two common shapes are the fink type and the fan type, as illustrated in Figure 5.13.

References/Further Reading

1 Chudley, R. and Greeno, R. (2008). *Building Construction Handbook*. Oxford: Butterworth-Heinemann.
2 Cooke, R. (2007). *Building in the 21st Century*. Oxford: Blackwell Publishing.

6

Introduction to Building Services

LEARNING OUTCOMES

1) Explain cold water supply, hot water supply, central heating and drainage.
2) Explain the integration of building services into the building design.

6.1   Introduction

A building would just be an assemblage of materials without the provision of building services. We need these services not only for comfort but also for our entertainment, convenience and communications. As building services involve the provision of pipes, cables and ducts, it is important that they are integrated into the building design in such a way that the building is not affected aesthetically. In this chapter, brief information on cold water supply, hot water supply, central heating and drainage will be provided; information on heat loss from buildings, electricity, light and fluid mechanics is provided in other chapters.

As the technical information on the topics discussed in this chapter is quite detailed, readers should refer to the references given at the end of the chapter.

6.2   Cold Water Supply

Water is essential for the survival of humans, animals and plants. Water is important for our health as all cells and organs in the human body depend on water for their functioning. It is such an important substance that its shortage in some countries is a cause for major concern for all nations. We need water, not just in our houses and commercial buildings, but also for several industrial processes such as steel making, nuclear power, hydro-electricity, building and civil engineering projects etc.

The chemical formula of water is H_2O, which shows that it is made of two atoms of hydrogen and one atom of oxygen. When it rains, some of the rainfall flows into rivers, streams and lakes and some percolates into the ground to form aquifers. As rain falls on the ground, the water gets contaminated due to gases and impurities in the atmosphere and the ground, and before water can be supplied to houses and other buildings, chlorination is carried out at treatment plants to kill bacterial microbes.

Water is supplied to homes via a network of pipes known as mains. On new installations, PVC pipes are used as water mains. The connection to the main is made using a ferrule, which is basically an isolating valve. The supply pipe from the water main to a dwelling consists of two sections – the communication pipe and the service pipe – as shown in Figure 6.1. The communication pipe is the

Construction Science and Materials, Second Edition. Surinder Singh Virdi.
© 2017 John Wiley & Sons Ltd. Published 2017 by John Wiley & Sons Ltd.
Companion website: www.wiley.com/go/virdiconstructionscience2e

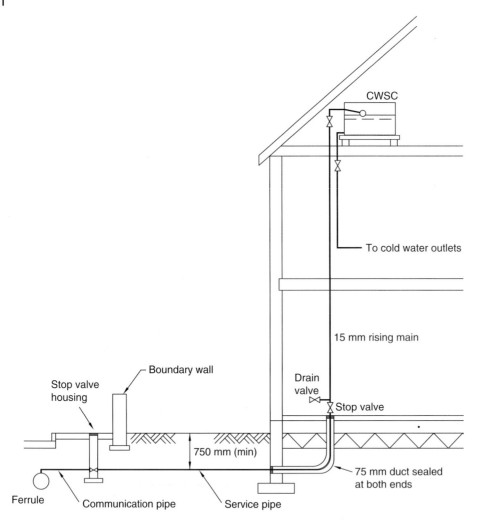

Figure 6.1 Water supply to a dwelling and indirect cold water system.

responsibility of the water authority whereas the service pipe is the responsibility of the house owner. When the service pipe enters a dwelling, protection is provided by encasing it in a 75 mm duct, so that it is not damaged if the external wall is affected by settlement. The water supply pipe terminates in the building with the provision of a drain valve and a stop valve, either as a separate or a combined fitting.

Within a house, water is supplied to the various outlets either from a direct or an indirect cold water system. In a direct cold water system, water is supplied to the outlets at mains pressure; therefore, for water storage, only a small-capacity cistern (100–150 litres) is required. This system requires less pipework.

In the indirect system (Figure 6.1) there is only one tap (at the sink) that supplies the drinking water from the mains supply. All other cold water taps are supplied water from the cistern; therefore, a 230–250 litre capacity cold water storage cistern (CWSC) is required. This system requires more pipework but provides a better reserve of water in cases where the mains supply fails or is cut off.

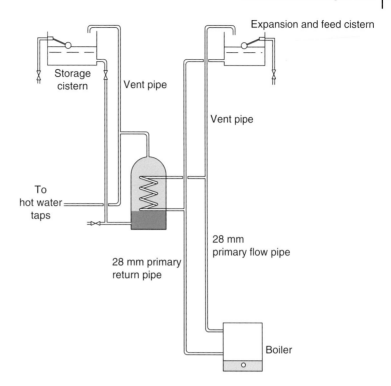

Figure 6.2 Hot water (indirect) system.

6.3 Hot Water Supply

A direct hot water supply is the most economical and simplest system. Hot water from the boiler rises by convection to the hot water cylinder and cooler water from the cylinder moves down into the boiler. Eventually, all of the water in the hot water cylinder is heated. When hot water is used in the kitchen or in the bathroom, it is replaced with cold water from the cold water storage cistern. This system cannot be incorporated into the central heating circuit. There is also the likelihood of scale build-up in hard water areas as large volumes of water are heated.

In an indirect system, the water contained in the hot water cylinder is heated indirectly through a heat exchanger. This is the most common domestic hot water system as it allows the boiler to be used for the central heating circuit (see Figure 6.2).

6.4 Central Heating Systems

There are several types of domestic central heating system, but here only the one-pipe and the two-pipe systems will be discussed. In the one-pipe system, there are no separate flow and return pipes, but just one pipe doing two functions. The system is easy to install, but there is a significant difference between the temperatures of the first and the last radiators. One way of alleviating this problem is to balance the radiators carefully.

The two-pipe system is far more efficient than the one-pipe system. Each radiator receives hot water at approximately the same temperature (about 70 °C) from the flow pipe, and the hot water leaving the radiator is returned to the boiler via the return pipe. The system includes a room thermostat

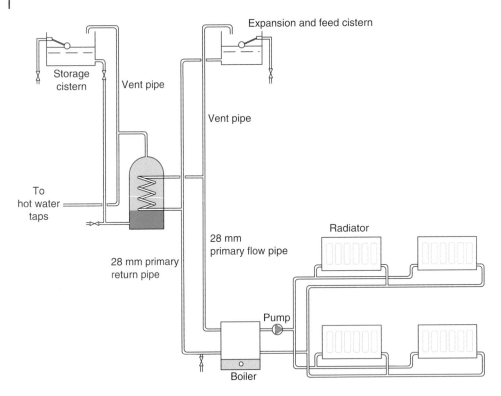

Figure 6.3 Two-pipe semi gravity system.

which switches off the pump when the desired room temperature is reached. Similarly, the temperature of the secondary water is also thermostatically controlled. Figure 6.3 shows the details of the two-pipe system.

6.5 Underfloor Heating Systems

Underfloor heating (UFH) is another method of heating a building. UFH systems are adaptable and can be installed in both new and old buildings with a wide range of floor coverings. There are generally two types of underfloor heating system: the **wet system** (uses water) and the **dry system** (uses electricity). Whichever system is used, the pipes/cables are entirely concealed in the floor, gently radiating heat evenly throughout the room.

The wet system consists of laying the pipework, 150 to 250 mm apart, on a thermally insulated bed, in a loop or coil formation covering the floor of the room (Figure 6.4). The pipes are connected to the boiler which circulates warm water and, due to the insulated bed, most of the heat is directed upwards into the room. Because the heat emitted is more uniformly distributed than a single radiator, the system can use water at a lower temperature, and hence it is more energy efficient.

An electric underfloor heating system consists of a series of electric wires installed beneath or within the flooring. The main reasons for installing electric underfloor heating are that the system is simple to install and does not add a lot of height to the floor. Although cheaper to install, it is more expensive to run as compared to a wet UFH system.

Figure 6.4 Wet underfloor heating system. Image courtesy of Warm Floors Ltd., UK.

6.6 Drainage Systems

All buildings have a network of pipework to take the waste solids and liquids, and rainwater, away from the property and discharge it harmlessly. The drainage of a dwelling can be divided into two systems: the below-ground system and the above-ground system.

6.6.1 Below-ground Drainage System

The purpose of a below-ground drainage system is to convey rainwater and waste water from toilets and kitchens to larger pipes known as sewers. A good drainage system should be simple and provided in such a way that:

- The drains are laid at suitable gradients;
- The flow of water (effluent) is self-cleansing.

For domestic buildings, the drains are connected to main sewers, which are controlled by the various water authorities. The main sewers will convey the effluent to the sewage treatment plant for processing, after which the treated water can be safely discharged. Three systems, i.e. combined, separate and partially separate, are used; the first two are described below.

Combined Drainage System

In some systems, particularly older systems, both waste water and surface (rain) water are drained into one sewer. The main drawback with this system is that surface water is relatively clean, but when mixed with waste water it has to be treated at a treatment plant, increasing the costs. It may also create health problems if the sewers are unable to cope with the volume of water flowing during storm conditions.

Separate Drainage System

In the separate system, the surface water is conveyed in surface water drains and sewers, and the waste water is conveyed in foul drains and sewers. A smaller volume of waste water needs treatment as the drains and the sewers are kept separate. The surface water can be drained into rivers, streams etc. without treatment. Sometimes, the volume of waste water may not be enough to ensure self-cleaning of the sewers, in which case, some surface water may be discharged into the sewers to improve the flow rate (a partially separate system). These systems are illustrated in Figure 6.5.

In the systems described above, it is important to provide access to the drains so that if any blockage occurs it can be cleaned. Access can be provided by including rodding eyes, access fittings, inspection chambers and manholes, whichever are appropriate. All soil and vent pipes should drain into an inspection chamber or manhole. Access should be provided at the head of each drain and also at changes in gradient, direction and pipe size.

When a drainage pipe passes through a wall or a trench-fill foundation, it must be protected against any damage that could occur due to settlement. The best solution is to provide a lintel and leave an all-around gap of 50 mm, as shown in Figure 6.6. More details can be found in the Building Regulations – Part H.

6.6.2 Above-ground Drainage System

There are a number of above-ground drainage systems but the most common for domestic dwellings is the primary ventilated stack system also known as a single-stack system (Figure 6.7). All appliances are closely grouped around the stack; therefore, an extra ventilating stack is not required. The location of a branch pipe in a stack should not cause cross-flow into another branch pipe; its size should always be at least the same diameter as the trap. Bends at the base of the discharge stack should have a radius of at least 200 mm; two 45° bends can be used as an alternative. A terminal mesh guard should be fitted at the top of the stack to prevent the possibility of birds nesting.

6.7 Integration of Services into Building Design

Modern buildings require many types of services such as heating, air-conditioning, electricity, fire protection etc. for the comfort, health, safety and security of people. Due to the unsightly appearance of service ducts, pipes and cables, provision is made in the structure of the building for accommodating them. The service routes extend both vertically and horizontally; vertically between floors (vertical risers) and horizontally to service the floors.

In a dwelling, water, gas and electrical services can be accommodated horizontally below the floor boards. For solid floors, the services can be housed in either a skirting duct or a floor duct provided in the concrete floor. In timber floors, the services may run at right angles to the joists, in which case, holes can be drilled and notches cut to accommodate cables and rigid pipes, respectively. There are rules for the position of holes and notches in the timber joists. The safest place to penetrate a joist is right through the centre, which, being halfway between the compressed top and the stretched bottom, is what is called the **neutral axis**. Even so, drilling should not be done at the mid-span of the joist, nor too close to the supporting ends, but in a zone between 0.25 and 0.4 times the span from the end. Where rigid copper pipes must run perpendicular to joists, they have to be laid in notches. These should be cut only in a zone between 0.07 and 0.25 times the span from the end. For a typical floor joist spanning 3.5 metres, this means the notches should be cut only between 250 mm and 880 mm from the wall. More importantly, the maximum safe depth for a notch is one-eighth the depth of the joist. So, for 200 mm joists, notches should be no deeper than 25 mm (see Figure 6.8(a)).

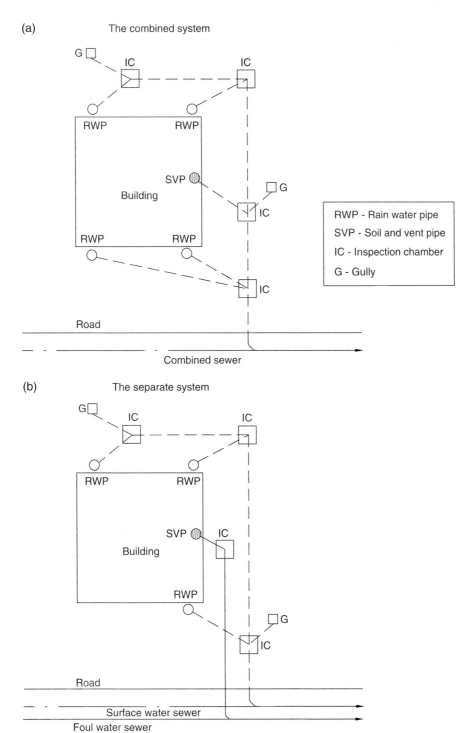

Figure 6.5 Below-ground drainage systems.

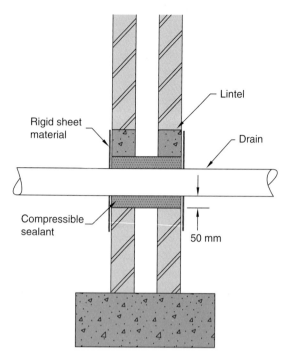

Figure 6.6 Protection of drains penetrating walls.

Rigid sheet material

Lintel

Drain

Compressible sealant

50 mm

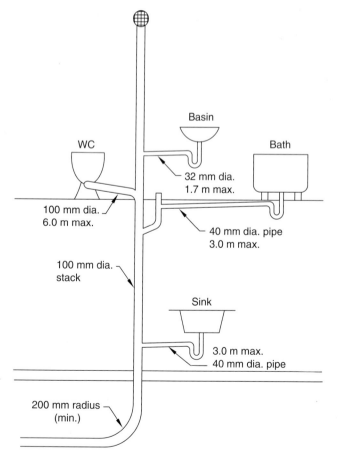

Figure 6.7 Primary ventilated stack system.

Basin

WC

Bath

32 mm dia.
1.7 m max.

100 mm dia.
6.0 m max.

40 mm dia. pipe
3.0 m max.

100 mm dia.
stack

Sink

3.0 m max.
40 mm dia. pipe

200 mm radius
(min.)

(a)

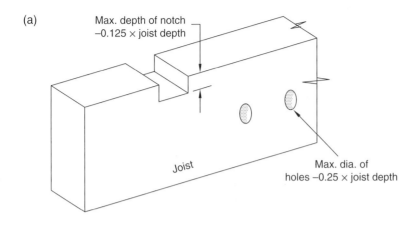

Max. depth of notch
−0.125 × joist depth

Joist

Max. dia. of
holes −0.25 × joist depth

(b)

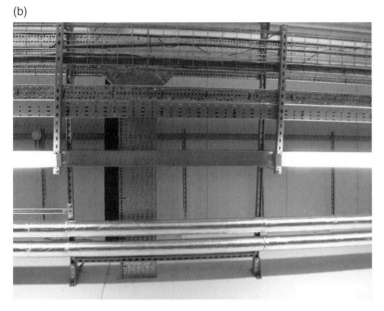

(c)

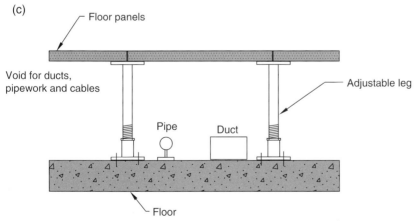

Floor panels

Void for ducts,
pipework and cables

Adjustable leg

Pipe

Duct

Floor

Figure 6.8 Integration of building services.

The designer may decide to provide an underfloor heating system. The system, whether it is a wet system or a dry system, can be integrated by providing the pipes/cables within the floor construction, as shown in Figure 6.4.

In commercial buildings, the services can be integrated horizontally either below or above a floor. Some common approaches for accommodating services between the soffit of a floor and the ceiling are:

- For shallow floor beams, the services can pass underneath the beams (Figure 6.8(b)).
- If a floor slab is supported by beams of moderate depth, some services can be accommodated between the beams, but ducts have still to pass beneath the beams.
- If the floor beams are deep enough, the services can pass through the beams so that the structure and the services occupy the same horizontal zone.

The services can also be integrated into the building design by accommodating them in raised floors. Using adjustable pedestals, a void 100 to 600 mm high can be created for the provision of services, as shown in Figure 6.8(c).

References/Further Reading

1 Hall, F. and Greeno, R. (2013). *Building Services Handbook*. Oxford: Routledge.
2 Oughton, D.R. and Hodkinson, S. (2005). *Heating and Air-conditioning of Buildings*. Oxford: Butterworth-Heinemann.
3 The Building Regulations (2010). *Approved Document H – Drainage and waste disposal*. London: Department of Communities and Local Government.

7

Thermal Energy 1

LEARNING OUTCOMES

1) Define temperature and heat.
2) Explain states of matter, latent heat and sensible heat.
3) Explain expansion and contraction of materials and solve numerical problems.
4) Explain the process of heat transfer and its practical applications.

7.1 Introduction

There are many forms of energy, thermal energy being one of them. Without thermal energy life on this planet would not be possible. We use heat daily in our homes for cooking, space heating and hot water. Many industrial processes also use heat energy, for example, the generation of electricity in thermal power stations.

Heat affects buildings and other structures in many ways, and to understand these processes and problems a study of the nature and effects of heat is imperative. In this chapter, the transmission of heat, the expansion and contraction of materials and thermal insulation of building elements will be discussed. But first it is necessary to define and discuss temperature, its measurement and the units of heat.

7.2 Temperature

The term 'temperature' is used to express the degree of hotness of a substance. It is measured by using a device called a thermometer. Of the different types of thermometer available, the mercury thermometer is the most widely used. This is basically a glass tube with one end closed and the other made to form a bulb, which acts as a mercury reservoir. The stem is marked into graduations called degrees, as shown in Figure 7.1. The figure also shows a digital thermometer which consists of a sensor known as a thermistor. Owing to a change in temperature, the resistance of the sensor changes; this is measured by the circuit in the device and converted into a temperature.

Construction Science and Materials, Second Edition. Surinder Singh Virdi.
© 2017 John Wiley & Sons Ltd. Published 2017 by John Wiley & Sons Ltd.
Companion website: www.wiley.com/go/virdiconstructionscience2e

(a)   (b)

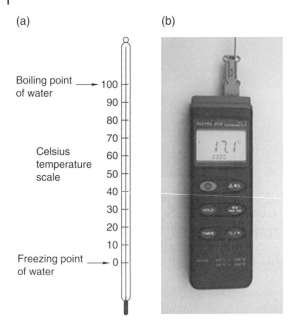

Figure 7.1 (a) Mercury and (b) electronic thermometers.

Boiling point — 100
of water

90

80

70

Celsius   60
temperature
scale   50

40

30

20

10

Freezing point — 0
of water

7.2.1 Temperature Scales

The Celsius Scale
The temperature of pure melting ice is zero degrees Celsius (written as 0 °C) and the temperature of steam from boiling water is 100 °C. Between 0 °C and 100 °C, the scale is divided into 100 equal divisions, each one being called a degree.

The Thermodynamic Temperature Scale
Charles's Law states that the volume of a fixed mass of gas at constant pressure is proportional to its temperature. If the temperature of the gas is lowered, the volume reduces as well. The temperature at which the volume will become zero is called the **absolute zero temperature**. This is the lowest temperature that can be attained theoretically, but is impossible practically, as the volume of a gas can never become zero. Using the absolute zero as the zero temperature, the thermodynamic temperature scale was devised. The scale is divided into units called kelvins denoted as K rather than °K. The size of one **kelvin** is the same as that of a degree Celsius.

$$0\,K = -273\,°C\,(\text{approximately})$$
$$273\,K = 0\,°C$$
$$373\,K = 100\,°C$$

7.3 Units of Heat

The quantity of heat required to raise the temperature of 1 gram of water through 1 °C is called a **calorie**. In the SI system, the unit of heat energy is the **joule** (J), which is defined as the work done when a force of 1 newton acts through a distance of 1 metre.

Power is defined as the rate of transfer of energy:

$$\text{Power} = \frac{\text{Energy transferred}}{\text{Time taken}}$$

The SI unit of power is the watt (W) and is equal to the transfer of one joule of energy per second (s):

$1\,\text{W} = 1\,\text{J/s}$

$1\,\text{kilowatt}\,(\text{kW}) = 1000\,\text{W}$

$1\,\text{megawatt} = 1000000\,\text{W}\left(\text{or }10^6\,\text{W}\right)$

7.4   States of Matter

Anything that occupies space, i.e. has volume, and has mass is called **matter**. Typical examples are: air, water, metals, wood, plaster, glass etc. There are three forms in which matter can occur: solid, liquid and gas. These forms are known as the states of matter. Water is one of the substances that we use daily, and the three states in which it exists are:

Solid – ice

Liquid – water

Gas – water vapour and steam

The properties of solids, liquids and gases depend on the forces between their molecules. The molecules of a solid are held together by strong forces of attraction. The movement of molecules past one another is not possible; therefore, the shape, volume and mass of a solid remain the same.

The molecules of a liquid are quite close, but the forces holding them together are not very strong. The molecules can move past one another, enabling a liquid to take the shape of the container in which it is stored; however, its volume remains constant.

A gas has definite mass, but the shape and volume can change, as the molecules of a gas are far apart. Compressing a gas, or changing its shape and volume, is quite easy.

7.4.1   Changes in the Physical State

In a physical change, the state may alter from solid to liquid or from liquid to gas and vice versa, but the basic structure of the substance remains unchanged. This is quite different from a chemical change, where the new material formed is different from the original substance. A chemical change is irreversible whereas a physical change is reversible. Ice changes into water on heating, but when the temperature of water is lowered below $0\,°\text{C}$, ice is formed, showing that this is a reversible change. The rusting of iron due to oxidation is a chemical change, as a new substance is formed, the properties of which are totally different from the parent material.

7.4.2   Experiment: The Physical States of Water

Apparatus and materials: Beaker, retort stand and clamp, thermometer, Bunsen burner, tripod and gauze, clock and ice.

Procedure:

1) Take some ice cubes and put them in the beaker.
2) Set up the apparatus as shown in Figure 7.2 and note the temperature; this should be below 0 °C.
3) Start heating the ice and record the temperature of the ice at regular intervals. The data may be recorded in a tabular form.
4) Continue to observe and note the temperature at regular intervals. The temperature will stay at 0 °C for some time while the ice is melting.
5) After all the ice has melted, the temperature of the water will start to rise until the water begins to boil. The temperature of boiling water, if it is pure, should be 100 °C. If there are impurities in the water, the boiling temperature may be slightly higher.
6) The temperature will stay at 100 °C as the water boils and evaporation takes place. The experiment may be terminated after a few minutes.

Results and discussion:

The data are plotted as temperature versus time, as shown in Figure 7.3; the graph is called the heating curve of ice. Ice is solid, and to change ice into water, heat energy is required. The temperature rises to 0 °C as the ice starts to melt and remains unchanged until all the ice has melted. This is shown as the 'ice and water' stage in Figure 7.3.

The amount of heat required to change ice into water, the temperature remaining unchanged at 0 °C, is known as the **latent heat** of ice. The latent heat of ice is 330 kJ/kg, which means that 330 kJ of heat energy are required to convert 1 kg of ice into water at 0 °C.

The temperature at which a substance begins to melt is known as its **melting point**; the melting point of ice is 0 °C. Further heating raises the temperature of the water (liquid state) steadily, as shown in the 'water' stage in Figure 7.3. Heat used in this way is called **sensible heat**. If the water is pure, it will begin to boil at 100 °C. The temperature remains unchanged as water is being converted into steam and vapour (the 'water and steam' stage). The reason for this is that heat is required to change boiling water into steam. This is called the latent heat of steam. Its value is 2250 kJ/kg, or, in other words, 2250 kJ of heat energy are required to convert 1 kg of boiling water into steam.

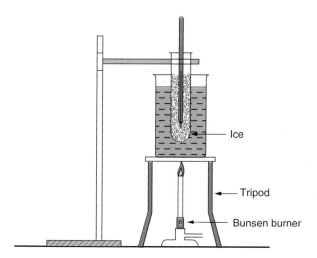

Figure 7.2

Figure 7.3 Heating curve for ice.

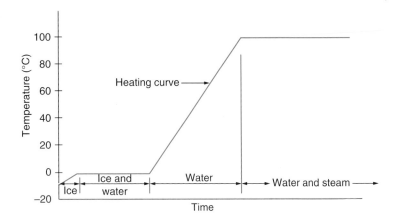

7.5 Expansion and Contraction of Solids

Buildings and civil engineering structures are affected by the expansion and contraction of materials as the surrounding temperature increases or decreases. Thermal expansion and contraction create stresses in buildings and other structures. Appropriate provisions are made in these structures so that the expansion or contraction takes place without creating any stress in the materials. Typical examples are the expansion and contraction joints in bridge decks and concrete roads. A detailed discussion is given in Section 7.5.3.

Thermal expansion or **thermal movement** is of three types: linear, superficial or cubical. Only linear expansion is explained here in detail.

7.5.1 Linear Expansion

If a bar of steel, or another metal, of length l is heated to raise its temperature from t_1 to t_2, there will be an increase in the length of the bar. The increase in length, Δl, is proportional to:

- The original length, l;
- The increase in temperature, Δt $(\Delta t = t_2 - t_1)$.

Combining the two above,

$$\Delta l \propto l \Delta t \ (\propto \text{ is the sign of proportionality})$$
$$\text{or, } \Delta l = \alpha l \Delta t$$
$$\text{therefore } \alpha = \frac{\Delta l}{l \Delta t}$$

where α, the constant of proportionality, is known as the **coefficient of linear expansion.**

The coefficient of linear expansion of a material is defined as the change in length per unit length per unit change in temperature. (Unit: per °C or mm/m °C.)

Table 7.1 gives the values of the coefficients of linear expansion of a range of common materials.

Table 7.1 Coefficients of linear expansion.

Material	α (per °C) (or per K)
Aluminium	0.000023 or (23×10^{-6})
Brass	0.000019 or (19×10^{-6})
Bricks	0.000007 or (7×10^{-6})
Concrete	0.000011 or (11×10^{-6})
Copper	0.000017 or (17×10^{-6})
Glass	0.0000085 or (8.5×10^{-6})
Invar	0.000001 or (1×10^{-6})
PVC	0.00007 or (7×10^{-5})
Steel	0.000012 or (12×10^{-6})

Example 7.1 A 2.0 m length of copper pipe is heated from 20 °C to 60 °C. Calculate the thermal movement if the coefficient of linear expansion of copper is 0.000017 per °C.

Solution:
Original length (l) = 2 m = 2000 mm
    Change in temperature (Δt) = 60 – 20 = 40 °C

$$\alpha = 0.000017$$
$$\alpha = \frac{\Delta l}{l\Delta t}$$

Change in length or thermal movement, $\Delta l = \alpha\, l\, \Delta t$
$$= 0.000017 \times 2000 \times 40$$
$$= \mathbf{1.36\,mm}$$

7.5.2 Experiment: Determination of Coefficient of Linear Expansion

Apparatus: Expansion apparatus as shown in Figure 7.4, metre rule, micrometer, Bunsen burner, tripod and gauze, thermometer (0–110 °C) and sample.
  Procedure:

1) The metals that can be used in this experiment are usually steel, aluminium, copper and brass, and are available as 500 mm (approximately) long tubes. Take the appropriate tube, enclose it in the insulation sleeve and pass the ends through metal blocks A, B and C.
2) Set up the apparatus as shown in Figure 7.4.
3) Measure the distance between the reference marks on blocks A and B; this is the original length of the tube (l).
4) Measure the distance between the blocks B and C (d_1) with a micrometer. Tighten the screws on blocks A and C.
5) Note down the room temperature (t_1 °C).
6) Light the burner and allow the steam to pass through the tube for 5 minutes when the water boils.

Figure 7.4 Determination of linear expansion.

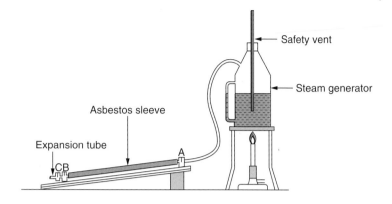

Table 7.2

Metal	Distance A-B (*l*) mm	Distance B-C (d$_1$) mm	Distance B-C (d$_2$) mm	Expansion (d$_2$−d$_1$)	Initial temperature (t$_1$)	Final temperature (t$_2$)	Temp. diff. (°C) (t$_2$−t$_1$)
Steel	502	9.0	9.45	0.45	20	100	100 − 20 = 80

7) Measure the distance between blocks B and C again (d$_2$). The difference between distances d$_2$ and d$_1$ gives the expansion of the tube.

8) Repeat the procedure with other metal tubes, if required. A table such as that shown in Table 7.2 may be used to record the test data.

$$\text{Coefficient of linear expansion of steel} = \frac{\Delta l}{l \times \Delta t}$$

$$= \frac{0.45}{502 \times 80} = \mathbf{0.000011 \text{ or } 11 \times 10^{-6} \text{ per } {}^{\circ}C}$$

7.5.3 Practical Examples of Expansion and Contraction

The expansion and contraction joints in concrete roads are shown in Figure 7.5. Concrete shrinks when it changes from a semi-liquid to a solid state and gains strength. If this factor is not taken care of, the concrete will crack haphazardly. By providing contraction joints, the cracking takes place only at predetermined places, i.e. at the contraction joints. To induce cracking at a predetermined place, the depth of the slab is reduced by incorporating fillets. Steel dowel bars are provided to strengthen the road slab and to maintain continuity.

In the UK, expansion joints in concrete roads are provided only if the construction takes place between 21 October and 21 April. Half the length of each dowel bar is coated with a debonding agent like bitumen, so that the concrete road is able to move as it expands. If the dowel bars are not debonded, concrete will grip them, making it difficult for the concrete to expand in a safe and satisfactory manner. The debonded sides of the dowel bars also carry compressible plugs, which absorb the movement of the concrete road.

In steel and concrete bridges, provision is made for one end to move in the event of expansion or contraction. In steel beams, the end is provided with rollers or rockers so that the bridge can expand

(a)

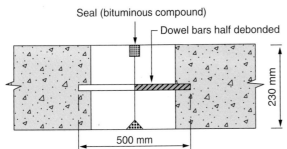

Seal (bituminous compound)

Dowel bars half debonded

230 mm

500 mm

(b)

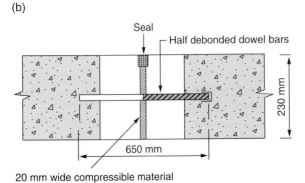

Seal

Half debonded dowel bars

230 mm

650 mm

20 mm wide compressible material

Figure 7.5 Joints in concrete roads: (a) a contraction joint; (b) an expansion joint.

(a)

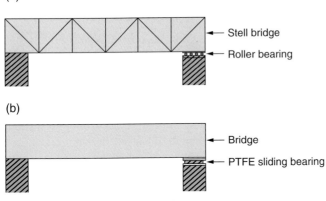

Stell bridge

Roller bearing

(b)

Bridge

PTFE sliding bearing

Figure 7.6 Bridge bearings.

or contract as the ambient temperature increases/decreases. In short-span concrete bridges, polyte-trafluroethylene (PTFE – a smooth plastic) bearings are provided to allow the bridge to expand or contract (see Figure 7.6).

Expansion joints are also provided in long brick walls; brick walls exceeding 15 m in length require expansion joints.

PVCu expands or contracts significantly due to changes in temperature. The use of this material in manufacturing doors, windows, conservatories and other components has become widespread in

Figure 7.7 Thermostat.

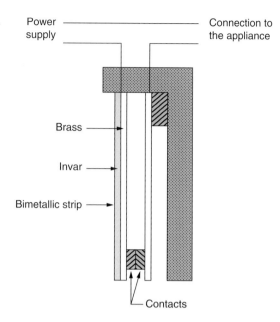

the UK. When old timber or metal windows are replaced by double-glazed PVCu windows, an all-around gap of 5 mm is provided to accommodate the expansion.

When strips of the same length, but of different metals, are heated through the same temperature, their expansions are not equal. For example, the expansion of brass is twenty times more than that of Invar. A strip made from these two substances, called a bimetallic strip, bends when it is heated. Owing to the higher coefficient of linear expansion of brass, the strip bends with the brass on the outside. This property is used to make a thermostat, which is a device used in boilers, heating systems, ovens etc. to control temperature. Figure 7.7 shows the principle of a thermostat which uses a bimetallic strip that can be used to control room temperature. When the two contacts touch, the current flows through the circuit and the heater is turned on. When the room temperature rises, the bimetallic strip expands and bends with the brass on the outside. The electrical contacts separate and the circuit to the heater is broken, thus switching off the heater. When the air temperature drops, the bimetallic strip straightens and the heater is switched on again as the electrical contacts touch. The thermostat may be set at different temperatures by turning the control knob.

7.6 Heat Transfer

Heat transfer from one substance to another, or from one point to another in a substance, may take place due to conduction, convection or radiation.

7.6.1 Conduction

If one end of a metal rod is put in a flame, the other end will become hot after some time. The process by which heat energy travels from one end of the rod to the other, without the molecules changing their positions, is known as **conduction**. Metals such as copper, silver, steel and aluminium allow the

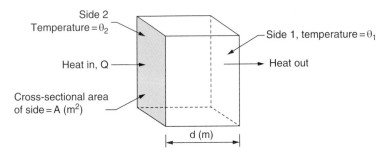

Figure 7.8

heat to pass through easily and quickly, and are called **good conductors** of heat. Non-metals and other substances like plastic, timber, cork, air and water are called bad conductors or **good insulators** as they do not conduct heat very well. Bad conductors have several applications in science and engineering, which are discussed in Chapter 8.

It is often necessary to compare the ability of various materials to conduct heat, and the property which is often used for this purpose is called the **thermal conductivity** or λ-value (also known as the k-value).

The thermal conductivity of a material is defined as a measure of the rate at which heat is conducted through it under specified conditions. The conditions are that the material should have a thickness of 1 m, a surface area of $1\,m^2$ and a temperature difference of $1\,^\circ C$ maintained between the opposite sides.

Consider a material of thickness d metres heated at one side, as shown in Figure 7.8. Heat comes out of the opposite side, assuming that heat loss from the other sides is prevented. The amount of heat, Q joules, which flows from side 2 to side 1 is:

- Directly proportional to the surface area (A) of side 2, as the greater the surface area, the higher the flow of heat.
- Directly proportional to the temperature difference, θ, between side 2 and side 1 ($\theta = \theta_2 - \theta_1$), as more heat will flow if the temperature difference between the two sides is large.
- Directly proportional to the time (t) for which the heat flows.
- Inversely proportional to the thickness (d) of the material. If the material is very thick, only a small amount of heat will flow through it and vice versa.

The amount of heat transferred will also depend on the nature of the material. Good conductors allow more heat to pass through as compared to bad conductors. Combining all the above factors, the amount of heat (Q) flowing through the material is proportional to:

$$\frac{tA(\theta_2 - \theta_1)}{d}$$

$$\text{or, } Q \propto \frac{tA(\theta_2 - \theta_1)}{d}$$

$$\frac{Q}{t} \propto \frac{A(\theta_2 - \theta_1)}{d}$$

In order to form an equation, the sign of proportionality is removed and a constant introduced. The constant used here is 'λ' (or 'k'):

$$\frac{Q}{t} = \frac{\lambda A(\theta_2 - \theta_1)}{d}$$

Table 7.3 Thermal conductivity of some building materials.

Material	Thermal conductivity (λ) W/m K
Aluminium	164
Brickwork (external)	0.84
Concrete – dense	1.5
Concrete – lightweight	0.57
Aerated concrete blocks	0.11
Insulation – expanded polystyrene board	0.033
– eco-wool	0.042
– glass fibre	0.04
– mineral wool batts	0.038
Plaster – dense	0.46
Plaster – lightweight	0.16
Plasterboard	0.16
Steel	43
Glass	1.02

Therefore:

$$\lambda \,(\text{or } k) = \frac{Qd}{tA(\theta_2 - \theta_1)}$$

λ in the above equation is known as the **thermal conductivity** of the material. A good conductor of heat will have a higher value of thermal conductivity than a bad conductor. Thermal conductivity values for a selection of materials are given in Table 7.3.

To work out the unit of thermal conductivity, the symbols in the above equation are replaced with appropriate units:

Quantity of heat, Q: joules (J)
Thickness of the material, d: metres (m)
Time during which heat flows, t: seconds (s)
Surface area of the material, A: m^2
Temperature difference, $\theta_2 - \theta_1$: °C or K

$$\lambda = \frac{Qd}{tA(\theta_2 - \theta_1)} \text{ becomes } \lambda = \frac{Jm}{sm^2 °C}$$
$$= \frac{Wm}{m^2 °C} \left(W = \frac{J}{s} \right)$$
$$= \frac{W}{m\,°C} \text{ (watts per metre per degree Celsius)}$$

If kelvin is used instead of °C, the unit will be $\dfrac{W}{mK}$

7.6.2 Experiment: To Compare the Thermal Conductivity of Metals

Apparatus: As shown in Figure 7.9, wax, Bunsen burner, beaker, tripod and gauze.
   Procedure:

1) Set up the apparatus as shown in Figure 7.9. The apparatus consists of a container and rods of steel, copper, aluminium, plastic and wood.
2) Apply a coating of molten wax over all the rods.
3) Boil some water in the beaker and pour the boiling water into the apparatus.
4) Start observing the rods and note the distance to which the wax has melted on various rods.

The distance to which the wax has melted is the highest in the case of the copper rod, meaning that the thermal conductivity of copper is higher than that of the other materials.

7.6.3 Convection

When water is heated in a vessel, it becomes hot after a few minutes. Although water is not a good conductor of heat, heat energy is still transferred from one point of the liquid to another. As water is heated, its particles at the bottom of the vessel become hot and expand as a result. Owing to expansion, the hot particles become lighter and rise. Their place is taken by cold particles from the upper section as they are slightly heavier than the hot particles. As the heating of the vessel continues, water will begin to boil after some time.

As water, like many other liquids, is a bad conductor of heat, heat is not transferred by conduction. Instead, heat is transferred from one part of the liquid to another by the movement of its molecules. This process is called **convection.**

Practical Examples of Convection
Like water, air is also a bad conductor of heat. Still, the heating of air in a room by a radiator or a heater can be done without any difficulty. The radiator or heater heats the surrounding air, which, as a result, becomes lighter and rises. Its place is taken by cooler air which is heavier; the upward movement of warm air and the downward movement of cooler air are called the convection currents. Figure 7.10 illustrates this process.

Figure 7.11 shows a domestic gravity hot water system. The boiler heats the water in the pipework. Hot water rises as it becomes less dense and lighter, and cold water descends to take its place. These convection currents will heat all the water in the cylinder.

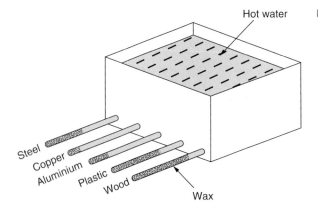

Hot water

Figure 7.9 Comparison of thermal conductivities.

Steel
Copper
Aluminium
Plastic
Wood

Wax

Figure 7.10 Transfer of heat by convection.

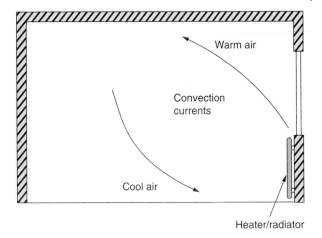

Figure 7.11 Convection currents in a hot water system.

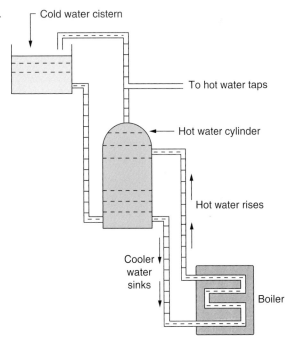

7.6.4 Radiation

The processes of conduction and convection need the presence of a substance for conveying heat from one point to another. **Radiation** is a process in which heat energy travels from one place to another without requiring a medium. Heat energy from the sun is received by us due to radiation as there is no medium (air) beyond the Earth's atmosphere. The radiant heat given off by the sun and other hot objects consists of infra-red rays with a wavelength between 750 nm (nanometres) and 0.001 mm.

A good radiator of heat is a good absorber as well, for example matt black objects. Shiny silver surfaces are bad radiators and absorbers, but good reflectors. This property is used in reflecting solar heat from flat felt roofing either by covering it with light-coloured stone chippings or painting it with a light-coloured paint.

Exercise 7.1

1 Convert the following temperatures into kelvin:
   A 10 °C.
   B 30 °C.
   C − 50 °C.

2 Convert the following temperatures into °C:
   A 295 K.
   B 315 K.
   C 343 K.

3 A 500 mm long aluminium tube was tested to determine the coefficient of linear expansion of aluminium. The following data were obtained from the test:

   Initial temperature of the tube $= 20\,°C$

   Final temperature of the tube $= 100\,°C$

   Expansion of the tube $= 0.91\,mm$

   Calculate the coefficient of linear expansion of aluminium.

4 A 5.0 m length of steel tube is heated from 10 °C to 40 °C. Calculate the thermal movement if the coefficient of linear expansion of steel is 12.0×10^{-6} per °C.

References/Further Reading

1 Chudley, R. and Greeno, R. (2008). *Building Construction Handbook*. Oxford: Butterworth-Heinemann.
2 CIBSE Guide A (2006). *Environmental Design*. London: Chartered Institution of Building Services Engineers.

8

Thermal Energy 2 (Including Humidity)

LEARNING OUTCOMES

1) Compare the thermal insulation values of different materials.
2) Explain thermal resistance and the coefficient of thermal transmittance.
3) Calculate the U-values of building elements.
4) Explain surface and interstitial condensation, and carry out calculations to determine the possibility of interstitial condensation in cavity walls.

8.1   Introduction

The basic concepts of heat transfer have been explained in Chapter 7. This chapter deals with one of the important aspects of our daily life, i.e. heat energy and its conservation. Heat energy is produced mainly from natural gas and electricity. Electricity is generated from several natural resources but only wind power, wave power and solar energy are renewable and do not create any pollution. They might have some impact on the environment, but the use of coal, water (hydro-electric projects) and nuclear fission/fusion can be more harmful to the environment. It is therefore important to conserve heat energy and reduce the demand for natural gas and electricity, which, in turn, will reduce the harm done to the environment.

In domestic buildings, offices, shops etc., heat loss in winter occurs due to the difference between the internal and the external air temperatures. The temperature inside a building is maintained at about 20 °C, but the external temperature fluctuates and could be sub-zero for several days/weeks. If a building is uninsulated and has single-glazed windows, then a large amount of heat is lost through its fabric (walls, roof, windows, floor and doors).

8.2   Thermal Insulation

Figure 8.1 shows the typical values of heat loss from an uninsulated house. About 70% of the heat loss occurs through the roof, walls and windows, which can be reduced by insulating the loft and the external walls, and by replacing the old timber/metal windows with PVCu double-glazed windows.

In new buildings, heat energy can be conserved by providing ground-floor insulation in addition to the above measures.

Construction Science and Materials, Second Edition. Surinder Singh Virdi.
© 2017 John Wiley & Sons Ltd. Published 2017 by John Wiley & Sons Ltd.
Companion website: www.wiley.com/go/virdiconstructionscience2e

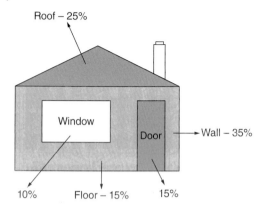

Roof – 25%

Window

Door

Wall – 35%

10%

Floor – 15%

15%

Figure 8.1 Typical heat loss for an uninsulated single-glazed house.

There are several insulation materials that can be used in a building, the most common being:

> **Flexible:** mineral wool and glass fibre quilts/batts;
> **Loose:** expanded polystyrene granules, vermiculite and mineral wool;
> **Rigid:** aerated concrete blocks, lightweight concrete and expanded polystyrene slabs.

Recently, eco-wool, which is made from recycled plastic bottles, has become popular, as its fibres do not cause any skin irritation or breathing problems.

Other insulating materials are wood, asbestos, sawdust and cork. Air, although not a material, is a very good insulator as its thermal conductivity is only 0.024 W/m K. For conservation of heat, we need to use materials that resist the flow of heat, i.e. have very low values of thermal conductivity. Table 8.1 shows the values of thermal conductivity for a range of construction materials.

There are several criteria that an insulating material has to meet. A good insulating material should be resistant to moisture and fungi, and should have adequate fire resistance. It should be available in different sizes and be harmless to humans and the environment. An experiment to compare the insulation values of different materials can be performed in the laboratory, as described in the next section.

8.2.1   Experiment: To Compare the Thermal Insulation Values of Expanded Polystyrene, Vermiculite, Mineral Wool, Glass Fibre and Cork

Apparatus: Five beakers, five thermometers, five test tubes with corks to hold thermometers, a large beaker, Bunsen burner, tripod and gauze, five stop-clocks and materials as mentioned.
Procedure:

1) Arrange the apparatus and the materials as shown in Figure 8.2. Boil some water in the large beaker and pour sufficient quantities into all the test tubes.
2) After about 5 minutes empty the test tubes and refill them with boiling water and replace the corks containing the thermometers.
3) Start to record the temperature of each tube, making sure that all five temperatures are the same.
4) Read the thermometers every minute initially, increasing the interval between readings gradually.
5) The experiment may be terminated after about 20 minutes. Table 8.2 may be used for recording the data.

Table 8.1 Thermal conductivity of some building materials.

Material	Thermal conductivity (λ) W/m K
Aluminium	164
Brickwork (external)	0.84
Concrete – dense	1.5
Concrete – lightweight	0.57
Aerated concrete blocks	0.11
Copper	386
Glass	1.02
Insulation – cork board	0.04
– expanded polystyrene board	0.033
– eco-wool	0.042
– glass fibre	0.04
– mineral wool batt	0.038
– woodwool slabs	0.09
Marble	2.0 to 2.9
Plaster – dense	0.46
Plaster – lightweight	0.16
Plasterboard	0.16
Steel	43
Timber/timber products	
– softwoods (fir, pine)	0.12
– hardwoods	0.16 to 0.23
– plywood	0.12

Figure 8.2

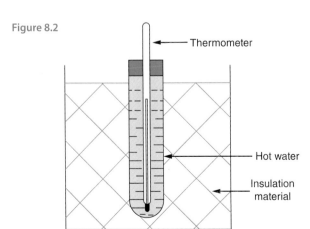

Table 8.2

Time (minutes)	Temperature (°C)				
	Expanded polystyrene	Vermiculite	Mineral wool	Glass fibre	Cork
0					
1					
2					
....					

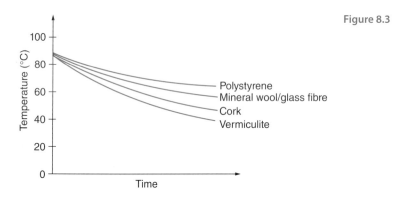

Figure 8.3

Results:

The temperatures of all the materials are plotted against time on the same graph, as shown in Figure 8.3. The graphs are called the cooling curves. The material that shows the lowest rate of cooling, i.e. the lowest drop in temperature, is the best insulator.

8.3   Heat Transmission

8.3.1   Thermal Conductivity

Thermal conductivity of a material (also known as the **λ-value** or **k-value**) is a measure of whether it will allow heat to pass through easily or offer resistance. The coefficient of thermal conductivity of a material, λ, is the amount of heat flowing (joules) per second across the opposite sides of a material 1 m thick, and of cross-sectional area 1 m^2, the sides being maintained at a temperature difference of 1 K.

$$\lambda = \frac{W \times m}{m^2 \times T} \left(\text{see Chapter 7, Section 7.6.1} \right)$$

Unit : W/m K or W m^{-1}K^{-1}.

8.3.2 Thermal Resistivity (r)

Thermal resistivity is the reciprocal of thermal conductivity. A high value of thermal resistivity will indicate that the material offers a high resistance to the passage of heat, and hence is a good insulator. Like thermal conductivity, thermal resistivity is also based on unit measurements of a material, i.e. 1 m thickness and 1 m^2 cross-sectional area.

$$r = \frac{1}{\lambda}$$

Unit : $m\,K/W$ or $m\,K\,W^{-1}$.

8.3.3 Thermal Resistance (R)

Thermal resistivity of a material is based on 1 m thickness, but most building materials are available in different sizes and can be used to produce components of various thicknesses. Thermal resistance, therefore, is based on the actual thickness (d) of a material.

$$R = rd$$
$$\text{or,} \quad R = \frac{d}{\lambda} \left(r = \frac{1}{\lambda} \right)$$

Unit : $m^2 K/W$ or $m^2 K\,W^{-1}$.

Depending on the constructional details of a component, its total thermal resistance may be made up of the material resistance, the surface resistances and the airspace resistance:

- **Material resistance:** Each material offers some resistance to the transmission of heat. Good conductors like copper have low resistance whereas bad conductors like expanded polystyrene have high resistance to the transmission of heat. The material resistance is directly proportional to the thickness and inversely proportional to the thermal conductivity:

$$\text{Material resistance} = \frac{d}{\lambda}, \text{where d is the thickness of the material in metres.}$$

- **Surface resistance:** Air inside and outside a building is always moving due to temperature variations, ventilation and other effects. The surfaces of all building materials have some irregularities which trap air; therefore, stationary layers of air are formed in close contact with the surfaces – internal as well as external. These stationary layers of air act like insulating layers due to the low thermal conductivity of air ($\lambda_{air} = 0.024\,W/m\,K$).
  The air movement inside a building is far less than the air movement outside a building; therefore, the internal surface resistance is greater than the external surface resistance. Typical values are given in Table 8.3.
- **Airspace resistance:** Inside an airspace, for example in a double-glazed window or a cavity wall, heat may be transmitted by conduction, convection or radiation, or any combination of the three. The thermal resistance of airspace is due to the low thermal conductivity of air. The airspace resistance depends on the width of the cavity and whether it is ventilated or unventilated, as shown in Table 8.3.

Table 8.3 Standard thermal resistances.

Type of resistance	Building element	Surface emissivity	Heat flow	Thermal resistance (m² K/W)
Inside surface	Walls	High	Horizontal	0.12
		Low		0.30
	Roofs (flat or pitched), floors/ceilings	High	Upward	0.10
		Low		0.22
	Floors/ceilings	High	Downward	0.15
		Low		0.56
Outside surface	Walls	High	Horizontal	0.06
		Low		0.04
	Roofs (flat or pitched)	High	Upward	0.05
		Low		0.05
Airspaces (unventilated)	5 mm wide	High	Horizontal/Upward	0.11
	20 mm or more	High	Horizontal/Upward	0.18

The total thermal resistance of an element depends on its construction. If there is no airspace in the element, then the total thermal resistance is the sum of the surface resistances and the material resistances. For example, in a solid brick wall, plastered on one side, the total thermal resistance (R_{total}) is given by:

$$R_{total} = R_{si} + R_{plaster} + R_{brickwork} + R_{so}$$

In an element that incorporates airspace, for example a double-glazed window, the total thermal resistance is given by:

$$R_{total} = R_{si} + R_{glass} + R_{airspace} + R_{glass} + R_{so}$$

where R_{si} = the resistance of the inside surface
R_{so} = the resistance of the outside surface
R_{glass} = the material resistance
$R_{airspace}$ = the airspace resistance

8.4   Thermal Transmittance

Thermal transmittance, or the **U-value,** indicates the amount of heat energy that will flow per second through one square metre of a building element when the temperature difference between the inside and outside surfaces is one kelvin.

The current Building Regulations specify the following U-values for dwelling houses. These values have come down gradually by about 40% over the last two decades. The lower the U-value of an element, the better it is at resisting the flow of heat. The U-values shown in Table 8.4, taken from the current Building Regulations, are the area-weighted average values.

Table 8.4

Building element	U-value (W/m² K)
External walls	0.35
Roofs	0.25
Floor	0.25

The U-value of an element/component can be determined from the following relationship:

$$U = \frac{1}{R_{total}}$$

The following table or the spreadsheet or the equation developed in Section 8.3.3 may be used to calculate the U-values:

Material/ layer	Thermal conductivity, λ	Thickness, d (m)	Thermal resistance, $R = \dfrac{d}{\lambda}$ (m² K/W)	U-value $U = \dfrac{1}{R_{total}}$ (W/m²K)
			$R_{total} =$	

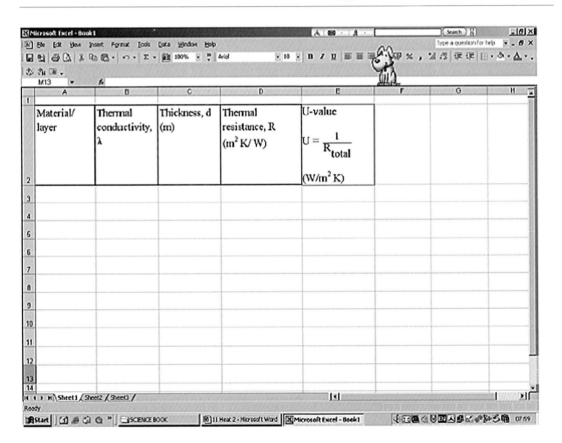

Example 8.1 Find the U-values of:

a) A single-glazed window.
b) A double-glazed window.

Given:
  Thickness of glass = 4 mm
  Thermal conductivity of glass = 1.02 W/m K
  Thermal resistance of surfaces: 0.12 (inside), 0.06 (outside) m^2 K/W (from Table 8.3)
  Thermal resistance of 20 mm wide airspace = 0.18 m^2 K/W (from Table 8.3)

Solution:
The first step is to draw the cross-section of the window (Figure 8.4) and list the layers of construction, including the surfaces, in a tabular form. The total thermal resistance of the component is determined, and its reciprocal calculated to obtain the U-value.

a) Single-glazed window

Material/layer	Thermal conductivity, λ (W/m K)	Thickness, d (m)	Thermal resistance, R (m^2 K/W)	U-value (W/m^2 K)
Inside surface			0.12	$U = \dfrac{1}{R_{total}}$
Glass	1.02	0.004	$\dfrac{d}{\lambda} = 0.004 / 1.02$ $= 0.0039$	
Outside surface			0.06	$= \dfrac{1}{0.1839}$
			$R_{total} = 0.1839$	$= 5.44$

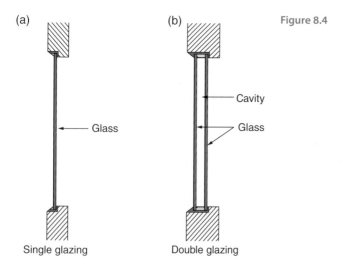

(a)   (b)   Figure 8.4

— Cavity

— Glass   Glass

Single glazing   Double glazing

b) Double-glazed window

Material/layer	Thermal conductivity, λ (W/m K)	Thickness, d (m)	Thermal resistance, R (m² K/W)	U-value (W/m² K)
Inside surface			**0.12**	$U = \dfrac{1}{R_{total}}$
Glass	1.02	0.004	0.004 / 1.02 = **0.0039**	
Cavity			**0.18**	
Glass	1.02	0.004	0.004 / 1.02 = **0.0039**	
Outside surface			**0.06** $R_{total} = 0.3678$	$= \dfrac{1}{0.3678}$ $= \textbf{2.72}$

Example 8.2

a) Calculate the U-value of a cavity wall with a half-brick-thick outer leaf, a 100 mm wide cavity without insulation, a 100 mm thick aerated concrete (Aer. Conc.) block inner leaf and a 12 mm thick layer of lightweight plaster.

b) Repeat the calculations if the cavity is provided with 100 mm thick mineral wool.

Thermal conductivities (W/m K):

> Aer. Conc. blocks = 0.11;
> lightweight plaster = 0.16;
> bricks = 0.84; mineral wool = 0.04

Surface resistances (m² K/W):

> inside surface = 0.12;
> outside surface = 0.06

Airspace resistance = 0.18 m² K/W

Solution:
Figure 8.5 shows the cross-section of the wall.

The question may be solved either electronically using a spreadsheet or by manual calculations. In this example, both methods are used: part (a) is solved by manual calculations and part (b) electronically.

In the spreadsheet, the formulae are entered in the first row and replicated for the other rows.

(a)

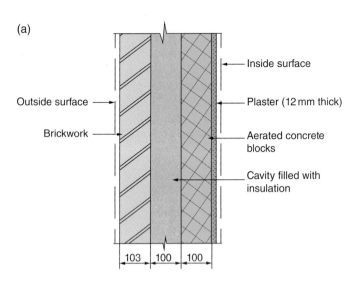

Figure 8.5 Cross-section of a cavity wall.

Outside surface →

Brickwork →

Inside surface

Plaster (12 mm thick)

Aerated concrete blocks

Cavity filled with insulation

103 · 100 · 100

Material/layer	Thermal conductivity, λ (W/m K)	Thickness, d (m)	Thermal resistance R (m² K/W)	U-value (W/m² K)
Inside surface	–	–	**0.12**	$U = \dfrac{1}{R_{total}}$
Lightweight plaster	0.16	0.012	0.012 / 0.16 = **0.075**	
Aer. Conc. blocks	0.11	0.100	0.100 / 0.11 = **0.909**	
Cavity	–	0.100	**0.18**	
Bricks	0.84	0.103	0.103 / 0.84 = **0.123**	
Outside surface	–	–	**0.06**	
			$R_{total} = 1.467$	$= \dfrac{1}{1.467}$ $= \mathbf{0.68}$

(b)

Material/ layer	Thermal conductivity, λ	Thickness, d (m)	Thermal resistance, R (m² K/ W)	U-value $U = \dfrac{1}{R_{total}}$ (W/m² K)		
Inside Surface			0.120			
Lightweight plaster	0.16	0.012	0.075			
Aer. Conc. Blocks	0.11	0.100	0.909			
Mineral wool	0.04	0.100	2.500			
Bricks	0.84	0.103	0.123			
Outside surface			0.060			
			3.787	0.26		
		R_{total}				

(The formulae used in this spreadsheet are shown in Appendix 1.)

Example 8.3   Calculate the U-value of an external wall of a timber-framed building shown in Figure 8.6.

Thermal conductivities of materials (W/m K):

> plasterboard = 0.16;
> mineral wool = 0.038;
> plywood = 0.12;
> bricks = 0.84

Surface resistances (m² K/W):

> inside surface = 0.12;
> outside surface = 0.06

Airspace resistance (m² K/W) = 0.18

Solution:

$$R_{total} = R_{si} + R_{plasterboard} + R_{mineralwool} + R_{plywood} + R_{airspace} + R_{bricks} + R_{so}$$

$$= 0.12 + \frac{0.0125}{0.16} + \frac{0.100}{0.038} + \frac{0.0095}{0.12} + 0.18 + \frac{0.103}{0.84} + 0.06$$

$$= 0.12 + 0.07813 + 2.6316 + 0.07917 + 0.18 + 0.1226 + 0.06$$

$$= 3.2715 \, m^2 K/W$$

$$\text{U-value} = \frac{1}{R_{total}} = \frac{1}{3.2715} = 0.31 \, W/m^2 K$$

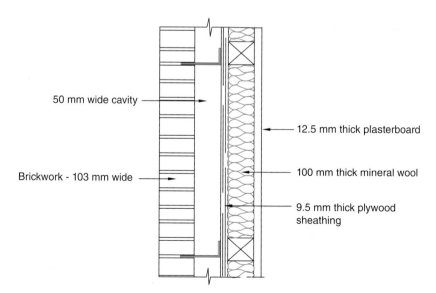

50 mm wide cavity

12.5 mm thick plasterboard

Brickwork - 103 mm wide

100 mm thick mineral wool

9.5 mm thick plywood sheathing

Figure 8.6

8.5   Heat Loss from Buildings

Heat is lost from buildings by conduction as well as convection. The former is also known as the heat loss through the fabric of the building, and the latter as the heat loss due to ventilation.

The **fabric heat loss** can be determined if we know the U-values of the building elements. We know the unit of U-value is:

$$U = W/m^2 \, K$$

Replacing m^2 and K with A (area of a component) and T (difference between the inside and outside temperatures) respectively, we have:

$$U = \frac{W}{A \times T}$$

After rearranging the terms, we have:

$$W \, (\text{rate of fabric heat loss}) = U \times A \times T$$

The **heat loss due to ventilation** occurs when the warm air of a room is replaced by fresh, but cooler, air from the outside. This change in air is necessary to replace the stale air with fresh air. This is usually achieved in two ways:

- **Uncontrolled ventilation:** Warm air escapes from the room when doors or windows are opened, or it escapes to the outside through gaps in the structure.
- **Controlled ventilation:** This is achieved by mechanical ventilation.

The fresh air needs to be heated for thermal comfort. The quantity of heat required to raise the temperature of $1 \, m^3$ of air by 1 K (or 1 °C) is known as the volumetric specific heat capacity of air. The volumetric specific heat capacity of air is 1212 $J/m^3 \, K$. The heating requirement due to the fresh air coming into a room is incorporated into the calculations as the number of air changes per hour. So, if two air changes per hour are required for a room, it means that two volumes of air have to be heated from the external temperature to the internal temperature every hour.

$$\text{Rate of heat loss due to ventilation} = \frac{c_v \times V \times N \times T}{3600}$$

where c_v = the volumetric specific heat capacity of air
  V = the volume of the room (m^3)
  N = the number of air changes per hour (Refer to Table 8.5 for recommended values of air change)
  T = the difference between the inside and outside air temperatures

The total rate of heat loss from a room is determined by adding the fabric heat loss and the ventilation heat loss.

$$\text{Total rate of heat loss} = \text{Fabric heat loss} + \text{Ventilation heat loss}$$

Table 8.5 Recommended allowances for air infiltration.

Building/room type	Air infiltration allowance/ air changes h^{-1}
Canteens and dining rooms	1
Dining and banqueting halls	0.5
Factories: 3000 m^3 to 10000 m^3	0.5 to 1
Over 10000 m^3	0.25 to 0.75
Houses, flats, hostels: Bathrooms	2
Bedrooms	0.5
Living rooms	1
Hotels: Bedrooms	1
Public rooms	1
Libraries: Reading rooms	0.5 to 0.7
Offices	1
Restaurants, cafes	1
Schools, colleges: Classrooms	2
Lecture rooms	1

(Reproduced from CIBSE Guide B: Heating, ventilating, air conditioning and refrigeration)

Example 8.4

a) Calculate the total rate of heat loss from the building shown in Figure 8.7. A temperature of 20 °C is to be maintained inside the building when the outside air temperature is 0 °C. Allow 2 air changes per hour; the volumetric specific heat capacity of air is 1212 J/m^3 K. The U-values (W/m^2 K) are:

Walls $= 0.35$; floor $= 0.25$; roof $= 0.25$; doors $= 2.2$; windows $= 1.9$.

The ground temperature is 5 °C.

b) Find the number of 900 × 600 mm high single convector radiators required for the building.

Solution:

a) The inside measurements should be used in this question:

$$\text{Walls: inside length} = 10.4 - 0.275 - 0.275 = 9.85 \, \text{m}$$
$$\text{Inside width} = 6.2 - 0.275 - 0.275 = 5.65 \, \text{m}$$
$$\text{Wall area} = 2 \times 9.85 \times 2.7 + 2 \times 5.65 \times 2.7 = 83.7 \, \text{m}^2$$
$$\text{Area of 2 doors} = 2 \times 0.84 \times 2.0 = 3.36 \, \text{m}^2$$
$$\text{Area of 6 windows} = 6 \times 1.5 \times 1.3 = 11.7 \, \text{m}^2$$
$$\text{Net wall area} = \text{Gross wall area} - \text{area of doors and windows}$$
$$= 83.7 - 3.36 - 11.7 = 68.64 \, \text{m}^2$$
$$\text{Floor area} = \text{roof area} = 9.850 \times 5.650 = 55.65 \, \text{m}^2$$

(a) 3D view (b) Plan

Doors: Two - 0.84 × 2.0 m high
Windows: Six - 1.5 × 1.3 m high

All walls are 275 mm thick

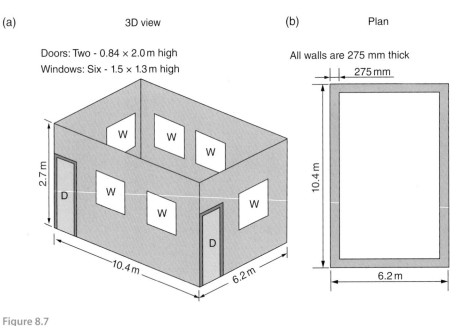

Figure 8.7

The rate of heat loss through the fabric of the building may be calculated by using the following table:

Element	U-value (W/m² K)	Area, A (m²)	Temperature difference, T (K)	Rate of heat loss = U × A ×T (watts)
Walls	0.35	68.64	20 – 0 = 20	480.48
Doors	2.2	3.36	20 – 0 = 20	147.84
Windows	1.9	11.7	20 – 0 = 20	444.60
Floor	0.25	55.65	20 – 5 = 15	208.69
Roof	0.25	55.65	20 – 0 = 20	278.25
				Total = 1559.86

The rate of heat loss due to ventilation is given by the formula:

$$\text{Heat loss} = \frac{c_v \times N \times V \times T}{3600}$$

Number of air changes, N = 2 per hour
Volume of the building, V = 9.85 × 5.65 × 2.7 = 150.26 m³
Difference between the inside and outside temperatures, T = 20 K

$$\text{Heat loss due to ventilation} = \frac{1212 \times 2 \times 150.26 \times 20}{3600} = 2023.50\,\text{W}$$

Total rate of heat loss = 1559.86 + 2023.50 = **3583.36 W** or **3583.36 joules/s**.

b) Total heat loss = 3583.36 W

   Try 900 × 600 mm high single convector radiators.

   Output of one radiator = 928 W (see information on radiators, Table 8.6)

   Number of radiators required = 3583.36 ÷ 928 = 3.86

   Provide four 900 × 600 mm high single convector radiators.

Table 8.6 Heat output of compact radiators.

Output based on a mean radiator water temperature of 70 °C and a room temperature of 20 °C

Nominal	Length	Single convector		Double panel plus		Double convector	
height	mm	btu/h	watts	btu/h	watts	btu/h	watts
400	500	1244	365	1066	521	2372	695
mm	600	1493	437	2132	625	2846	834
	700	1742	510	2487	729	3321	973
	800	1990	583	2842	833	3795	1112
	900	2239	656	3198	937	4270	1251
	1000	2488	729	3553	1041	4744	1390
	1100	2737	802	3908	1145	5218	1529
	1200	2986	875	4264	1249	5693	1668
	1400	3483	1021	4974	1457	6642	1946
	1600	3981	1166	5685	1666	7591	2224
	1800	4479	1312	6395	1874	8539	2502
	2000	4976	1458	...	...	9488	2780
500	300	906	266	...	...	1702	499
mm	400	1208	354	...	...	2269	665
	500	1510	443	2114	620	2836	831
	600	1812	531	2537	743	3403	997
	700	2114	620	2960	867	3971	1163
	800	2416	708	3383	991	4538	1330
	900	2718	797	3806	1115	5105	1496
	1000	3021	885	4229	1239	5672	1662
	1100	3323	974	4652	1363	6240	1828
	1200	3625	1062	5074	1487	6807	1994
	1400	4229	1239	5920	1735	7941	2327
	1600	4833	1416	6766	1982	9076	2659
	1800	5437	1593	...	...	10210	2992

(*Continued*)

Table 8.6 (*Continued*)

Output based on a mean radiator water temperature of 70 °C and a room temperature of 20 °C

Nominal	Length	Single convector		Double panel plus		Double convector	
height	mm	btu/h	watts	btu/h	watts	btu/h	watts
	2000	6041	1770	...	...	11345	3324
600	300	1056	309	...	...	1960	574
mm	400	1408	412	...	...	2613	766
	500	1759	516	2432	713	3266	957
	600	2111	619	2918	855	3919	1148
	700	2463	722	3404	998	4573	1340
	800	2815	825	3891	1140	5226	1531
	900	3167	928	4377	1283	5879	1723
	1000	3519	1031	4864	1425	6532	1914
	1100	3871	1134	5350	1568	7186	2105
	1200	4223	1237	5836	1710	7839	2297
	1400	4926	1443	6809	1995	9145	2680
	1600	5630	1650	7782	2280	10452	3062
	1800	6334	1856	...	...	11758	3445
	2000	7038	2062	...	...	13065	3828

(Extract from Quinn Radiators)

8.6 Temperature Drop Through Materials

If the two parallel faces of a uniform material are kept at different temperatures, then by using appropriate instruments it can be found that the temperature drops/increases uniformly through the material. Consider a cavity wall. If U is its U-value and T is the difference between the inside and outside temperatures, then

$$\text{Rate of heat loss} = U \times A \times T = U \times T \,(\text{assuming surface area}, A = 1\,\text{m}^2)$$

The heat loss takes place through each layer of construction. Considering each layer separately, its $\text{U-value} = \dfrac{1}{R_{layer}}$, where R_{layer} is the thermal resistance.

Therefore, for each layer, the temperature difference across its parallel faces (ΔT) can be determined from the equation:

$$U \times T = \frac{1}{R_{layer}} \times \Delta T$$

Transposing,

$$\Delta T = U \times T \times R_{layer}$$
$$= \frac{1}{R_{total}} \times T \times R_{layer}$$
$$= \frac{R_{layer}}{R_{total}} \times T$$

where R_{total} is the total thermal resistance of a wall.

Example 8.5   Find the temperature drop across each layer of the cavity wall dealt with in Example 8.2b and draw the structural temperature gradient. The inside and the outside temperatures are 20 °C and 0 °C, respectively.

Solution:
Temperature difference, T = 20 – 0 = 20 °C
    The calculations may be made in a tabular form, as follows:

Layer	Thickness, d (m)	Thermal conductivity λ (W/m K)	Thermal resistance of layer (R_{layer}) (m² K/W)	Temperature drop across layer $\frac{R_{layer}}{R_{total}} \times T$	Boundary temperature (°C)
					20
Inside surface	–	–	0.12	$\frac{0.12}{3.787} \times 20 = 0.6$	20 – 0.6 = 19.4
Plaster	0.012	0.16	0.012/0.16 = 0.075	$\frac{0.075}{3.787} \times 20 = 0.4$	19.4 – 0.4 = 19.0
Blocks	0.100	0.11	0.100/0.11 = 0.909	$\frac{0.909}{3.787} \times 20 = 4.8$	19.0 – 4.8 = 14.2
Insulation	0.100	0.04	0.100/0.04 = 2.5	$\frac{2.5}{3.787} \times 20 = 13.2$	14.2 – 13.2 = 1.0
Bricks	0.103	0.84	0.103/0.84 = 0.123	$\frac{0.123}{3.787} \times 20 = 0.7$	1.0 – 0.7 = 0.3
Outside surface	–	–	0.06	$\frac{0.06}{3.787} \times 20 = 0.3$	0.3 – 0.3 = 0
			Total = 3.787		

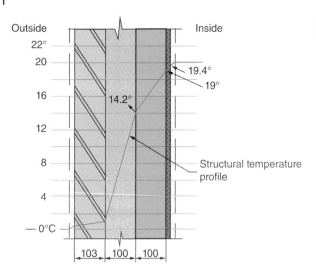

Figure 8.8 Structural temperature profile (or gradient).

The boundary temperatures are plotted to produce the structural temperature profile, as shown in Figure 8.8.

8.7 Humidity

Three-quarters of the Earth's surface consists of water in the form of oceans, seas, lakes and rivers. Evaporation of water occurs continuously, making the air humid and eventually leading to rain, snow, sleet etc. According to kinetic theory, molecules of water are in constant random motion. Sometimes there are collisions between molecules, causing some to gain enough energy to escape from the water surface into the atmosphere. As the water molecule enters the atmosphere, it becomes water vapour (gas state), well below the boiling temperature of water.

Air inside a building always has some moisture in the form of water vapour, which results from cooking, washing and breathing by the occupants. The amount of water vapour in the air is known as the **humidity**, and depends on the air temperature. It is important to know how close the humidity of an air sample is to the saturation point. This is known as the **relative humidity** (R.H.), and is expressed as:

$$\text{R.H.} = \frac{\text{Mass of water vapour in the air}}{\text{Mass of water vapour in the air if it were saturated at the same temperature}} \times 100$$

The relative humidity inside buildings should be between 40% and 60% for human comfort. If it is above 60%, the occupants don't feel comfortable as the perspiration from their bodies evaporates very slowly. The human body produces heat energy from food to maintain the body temperature at 37 °C. As the heat energy is produced continuously, there must be some process for the heat energy to escape so that the body temperature remains constant at 37 °C. Much of this heat is carried away in water vapour resulting from breathing and sweating. When the ambient temperature is high and the air is humid, the evaporation of perspiration is very slow to keep the body comfortable.

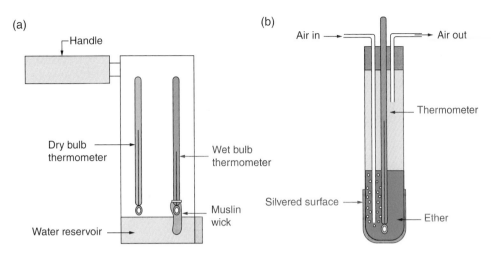

Figure 8.9 (a) Whirling hygrometer; (b) Regnault's hygrometer.

8.7.1 Measurement of Relative Humidity

Several methods are available for measuring relative humidity. The instrument used for this purpose is known as a **hygrometer**. Many types of hygrometers are available, for example:

- Wet and dry bulb hygrometer;
- Whirling hygrometer;
- The Regnault hygrometer;
- Paper or hair hygrometer;
- Electronic hygrometer.

The **wet and dry bulb hygrometer** and the **whirling hygrometer** (Figure 8.9(a)) work on the same principle. The whirling hygrometer is rotated manually to speed up the evaporation of water. There are two thermometers: the dry bulb thermometer and the wet bulb thermometer.

The dry bulb thermometer is exposed and used for recording the normal air temperature. The other thermometer is known as the wet bulb thermometer as its end, wrapped in muslin, dips into a water reservoir. As water evaporates from the muslin, it causes cooling and hence the temperature recorded by the wet bulb thermometer is lower than that recorded by the dry bulb thermometer. The readings from the two thermometers and Table 8.7 are used to find the relative humidity of the air.

The **Regnault hygrometer** consists of a glass tube, the lower end of which has a silvered surface. Other features of the apparatus are shown in Figure 8.9(b). Air is blown through liquid ether, causing it to evaporate. As ether evaporates, the latent heat required for evaporation is extracted from the surroundings. The tube and the air surrounding it are cooled down. Eventually, the air immediately around the bottom of the tube becomes saturated and dew is formed on the silvered surface. The temperature is noted and the tube is allowed to warm up by stopping the passage of air. The temperature is noted again when the dew disappears. The average of the two temperatures is the dew point. The relative humidity is found by obtaining the saturated vapour pressures (SVP) from tables and substituting in the formula:

$$\text{Relative humidity} = \frac{\text{SVP of the water vapour at the dew point}}{\text{SVP of water vapour at the air temperature}} \times 100$$

Table 8.7 Relative humidity (%).

Dry bulb temperature reading (°C)	Difference between dry and wet bulb readings (°C)									
	1	2	3	4	5	6	7	8	9	10
10	87	74	62	50	38	27	16	5		
12	88	76	65	54	43	32	22	12	2	
14	89	78	67	57	47	37	27	18	9	1
16	89	79	69	59	50	41	32	24	15	7
18	90	80	71	62	53	45	36	28	21	13
20	91	81	73	64	56	48	40	32	25	18
22	91	82	74	66	58	50	43	36	29	23
24	91	83	75	68	60	53	46	39	33	27
26	92	84	76	69	62	55	49	42	36	31

Figure 8.10 Electronic hygrometer.

Paper or hair hygrometers work on the principle that paper and hair increase in length when damp, and vice versa. This type of instrument is not very accurate because neither paper nor hair is perfectly elastic, i.e. they are permanently stretched eventually.

Electronic hygrometers (Figure 8.10) can measure the humidity of air and give a digital read-out. As the humidity of the air changes, the sensor responds to it and results in a change of static charge stored in the capacitor. This is displayed as relative humidity.

Example 8.6   Use Table 8.7 to determine the relative humidity of air for the following data:

Wet bulb temperature $= 14\,^{\circ}C$
Dry bulb temperature $= 20\,^{\circ}C$

Solution:
Difference between dry bulb and wet bulb temperatures $= 20 - 14 = 6\,^{\circ}C$.

Refer to Table 8.7. The relative humidity that corresponds to the dry bulb temperature of $20\,^{\circ}C$ and a temperature difference of $6\,^{\circ}C$ is the answer, i.e. **48%**.

8.8   Condensation

Condensation is said to occur if water vapour in the air changes its state to become water, i.e. it changes from a gas to a liquid. Air inside a building always has some moisture in the form of water vapour. The capacity of air to hold moisture depends on its temperature.

Water vapour exerts pressure whether the air is unsaturated or saturated with moisture. When the air is saturated, the pressure exerted by water vapour is known as the saturated vapour pressure, measured in N/m^2 or pascals (Pa). Table 8.8 gives the values of the SVP against a range of temperatures. An important conclusion that can be drawn from Table 8.8 is that if a sample of air is saturated at a certain temperature, it can be made unsaturated by increasing its temperature, and vice versa. For example, if a sample of air is saturated at $14\,^{\circ}C$, the SVP will be 1598 Pa. If the temperature of that air sample is raised to $20\,^{\circ}C$, more water vapour will be required as the SVP at $20\,^{\circ}C$ is 2337 Pa. In other words, the capacity of air to hold moisture increases as its temperature is raised, and vice versa.

The air inside a building is usually warm and unsaturated but the surfaces of window panes and external walls are cold, especially during winter. When warm air comes into contact with a cold surface, its temperature drops and its capacity to hold moisture decreases. As the external temperature continues to drop, the capacity of air to hold moisture keeps on decreasing, and eventually the air becomes saturated, with some moisture condensing and forming droplets of water on the cold surfaces.

The temperature at which water vapour starts to condense is called the **dew point**. (Unit: $^{\circ}C$ or K.)

This may occur as condensation on the wall and window surfaces (surface condensation) and/or within the materials of construction (interstitial condensation), as shown in Figure 8.11. Surface

Table 8.8 Saturated vapour pressure.

Temperature	SVP (Pa)	Temperature	SVP (Pa)
0	610	16	1818
2	705	18	2063
4	813	20	2337
6	935	25	3166
8	1072	30	4242
10	1227	40	7375
12	1402	50	12340
14	1598		

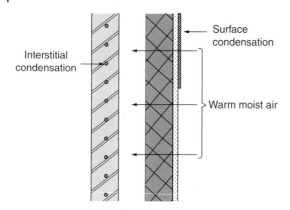

Figure 8.11 Condensation.

condensation can cause the deterioration of paint, wood, plaster and wallpaper. Interstitial condensation reduces the effectiveness of aerated concrete blocks and insulation materials, like mineral wool, in a cavity wall. This is due to the higher thermal conductivity of water (0.57) as compared to those of lightweight concrete blocks and mineral wool. A detailed discussion on interstitial condensation is given in Section 8.8.3.

8.8.1 The Psychrometric Chart

The psychrometric chart (Figure 8.12) shows the graphical interaction between the relative humidity of air, its temperature and dew point. The x-axis shows the dry bulb temperature, which is considered to correspond to the air temperature. The curves showing relative humidity of 0%, 10%, 20% etc. start from the y-axis on the left and slope upward. The uppermost curve represents a relative humidity of 100%. The y-axis on the right represents the moisture content of air and the vapour pressure. The wet bulb temperature, which is associated with moist air, is represented by the diagonal lines.

8.8.2 Prevention of Surface Condensation

It has already been discussed in Section 8.8 that the capacity of air to hold moisture decreases as the air temperature is lowered. When the air temperature is equal to its dew point, the air becomes saturated and the excess moisture is deposited on cold surfaces such as the inside of external walls and windows. Surface condensation may create several problems like mould growth, structural damage, deterioration of surface finishes and effects on the occupants' health. These problems can be reduced by taking one or more of the following actions:

- **Temperature:** Warm air can hold more moisture than cold air; therefore, by providing adequate heating, the problem of condensation can be solved. For condensation-free surfaces, it is important to ensure that the air temperature is above its dew point.
- **Ventilation:** The temperature of outside air is usually lower than the inside air temperature, hence its moisture content is lower than that of inside air. By providing adequate ventilation, the moisture-laden air from a building is carried outside and fresh air, containing much less moisture, is allowed to flow inside a building.
- **Insulation:** By providing adequate insulation to the building elements, we can make sure that the inside surfaces do not cool down very quickly when the heating stops. Typical examples of where the inside surfaces can cool down very quickly are uninsulated external solid brick walls and

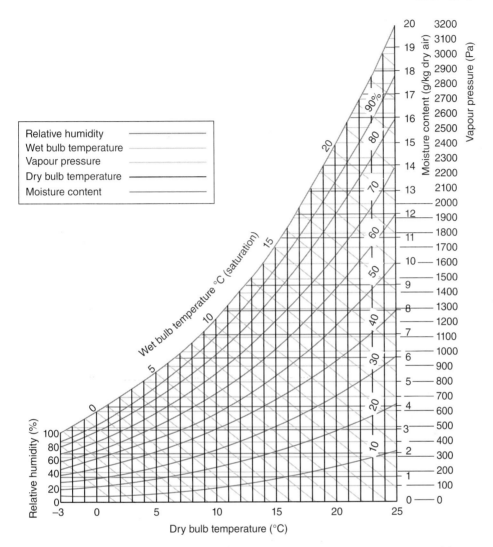

Figure 8.12 The psychrometric chart. (See the colour plate section for a full-colour version of this image.)

single-glazed windows. By replacing single-glazed windows with PVCu double-glazed windows, the problem of condensation on the window glass can be reduced. The air inside the double-glazed units acts as insulation. A similar effect can be achieved by providing adequate insulation to an uninsulated external solid brick wall.

Example 8.7 The temperature of air in a building is 19 °C and its relative humidity is 70%. Use the psychrometric chart and find:

a) The moisture content of air and the vapour pressure.
b) The relative humidity of air if the temperature is raised to 23 °C.
c) The temperature at which condensation will occur (the dew point).

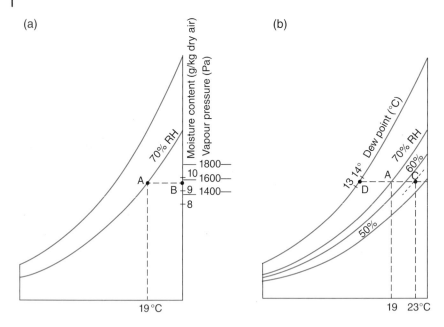

(a)   (b)

Figure 8.13

Solution:

a) On the psychrometric chart, the point where the vertical line from 19 °C (*x*-axis) meets the 70% R.H. curve is located (point A). A horizontal line is drawn from point A to meet the *y*-axis on the right at point B (Figure 8.13(a)). The values of the moisture content of the air and the vapour pressure are read off the vertical scales:

Moisture content $= 9.7$ g/kg
Vapour pressure $= 1560$ Pa

b) From point A draw a horizontal line to meet the 23 °C vertical line at point C (Figure 8.13(b)). The R.H. of the air at point C is 55% (approximately).

c) From point A draw a horizontal line to meet the 100% R.H. curve at point D. The dew point is 13.6 °C (Figure 8.13(b)).

8.8.3   Interstitial Condensation

Most building materials, with the exception of metals, are permeable to water vapour. Bricks, blocks, plasterboard and other materials will allow some amount of vapour to pass through. If the temperature of a material falls below the dew point of water vapour, condensation will occur. Condensation occurring within a building element is called **interstitial condensation**.

The rate at which water vapour passes through a material depends on the difference between the inside and outside vapour pressures. Internal and external surfaces and highly porous materials have a negligible resistance to the flow of water vapour; other materials show some resistance, as shown in Table 8.9.

Table 8.9 Typical values of vapour resistivity/resistance.

Material	Vapour resistivity (r_v) (MN s/g m)
Brickwork	25 to 100
Concrete	30 to 100
Expanded polystyrene	100 to 600
Mineral wool (insulation)	5
Plaster	60
Plasterboard	45 to 60
Plywood	1500 to 6000
Stone	150 to 450
Timber	45 to 75
Membrane	**Vapour resistance (MN s/g)**
Aluminium foil	4000
Gloss paint	7 to 40
Polyethylene sheet	110 to 120

The vapour resistance of a material (R_v), which measures its effectiveness in resisting the flow of water vapour, can be determined by multiplying its thickness (d) by its vapour resistivity (r_v):

$$R_v = r_v \times d$$

where d is the thickness in metres.

The overall resistance of a component to vapour flow may be obtained by adding the vapour resistances of all the constituent parts. The following sequence of calculations may be used to estimate the risk of interstitial condensation. In the first part, the temperatures across each layer of construction are calculated and, in the second part, the dew points are determined:

1) Calculate the thermal resistance of each material used in the construction of the element and hence find the total thermal resistance.
2) Find the difference between the internal and external temperatures (T). Starting from the inside, calculate the temperature drop (T_{drop}) across each material/layer of construction:

$$\text{Temperature drop across a layer} = \frac{\text{Thermal resistance of the layer} \left(R_{layer}\right)}{\text{Total thermal resistance} \left(R_{total}\right)} \times T$$

$$\text{or} \quad T_{drop} = \frac{R_{Layer}}{R_{Total}} \times T$$

where T is the difference between the inside and outside temperatures.
3) Calculate the boundary temperatures.
4) Draw a cross-section of the element and plot the points showing temperature drop across the various materials of construction. Join the points by straight lines to produce the structural temperature profile. This completes the first part of the calculations.
5) From the psychrometric chart, find the inside and outside vapour pressures, and the difference between them.

6) The water vapour flows from the high-pressure area towards the low-pressure area. Use the formula below to calculate the vapour pressure drop (P_{drop}) across each layer of construction:

$$\text{Vapour pressure drop across a layer} = \frac{\text{Vapour resistance of a layer }(R_v)}{\text{Total vapour resistance of element }(R_{v(total)})} \times P$$

or, $\quad P_{drop} = \dfrac{R_v}{R_{v(total)}} \times P$

where P is the difference between the inside and outside vapour pressures.

7) Calculate the vapour pressure at the boundary of each layer and find the dew points corresponding to these vapour pressures from the psychrometric chart.

8) Join all dew points by straight lines to obtain the dew point profile.

9) There is a risk of interstitial condensation if the structural temperature profile falls below the dew point profile.

The above procedure is followed in Example 8.8.

Example 8.8 A 300 mm thick (actual thickness is 303 mm) external wall is constructed of bricks, aerated concrete blocks, lightweight plaster and mineral wool. The inside air is at 22 °C and 50% relative humidity; the outside air is at 0 °C and 100% relative humidity. The properties of the materials and their thicknesses are:

Material	Thickness (d) mm	λ – value W/mK	Vapour resistivity (r_v) MNs/qm
Plaster	12	0.18	60
Aer. Conc. blocks	100	0.11	50
Mineral wool	100	0.04	5
Bricks	103	0.84	50

Determine if there is a risk of interstitial condensation.

Solution:
Difference between the inside and outside temperatures (T) = 22 – 0 = 22 °C

Material or layer	Thickness, d (m)	Thermal conductivity, λ (W/m K)	Thermal resistance (m² K/W)	Temperature drop (°C)	Boundary temperature (°C)
			$\dfrac{d}{\lambda}$	$\dfrac{R_{Layer}}{R_{Total}} \times T$	
					22
Inside surface	–	–	0.12	$\dfrac{0.12}{3.779} \times 22 = 0.7$	22 – 0.7 = 21.3

Plaster	0.012	0.18	= 0.012/0.18 = 0.067	$\frac{0.067}{3.779} \times 22 = 0.4$	21.3 – 0.4 = 20.9
Aer. Conc. blocks	0.100	0.11	= 0.100/0.11 = 0.909	$\frac{0.909}{3.779} \times 22 = 5.3$	20.9 – 5.3 = 15.6
Mineral wool	0.100	0.04	= 0.100/0.04 = 2.5	$\frac{2.5}{3.779} \times 22 = 14.6$	15.6 – 14.6 = 1.0
Brickwork	0.103	0.84	= 0.103/0.84 = 0.123	$\frac{0.123}{3.779} \times 22 = 0.7$	1.0 – 0.7 = 0.3
Outside surface	–	–	0.06	$\frac{0.06}{3.779} \times 22 = 0.3$	0.3 – 0.3 = 0

Total thermal resistance R_{total} = 3.779

Figure 8.14

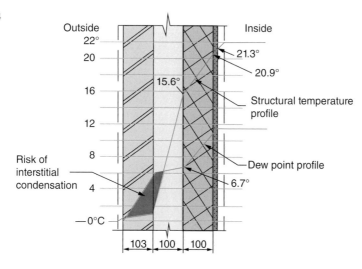

The cross-section of the wall is drawn to a suitable scale and the boundary temperatures are plotted (Figure 8.14). The points are joined by straight lines to produce the structural temperature profile. This completes the first part of the solution. In the second part, the vapour pressures at the boundary of materials are determined, which are then used to find the dew points from the psychrometric chart.

From the psychrometric chart:

Inside vapour pressure = 1340 Pa (air temperature = 22 °C; R.H. = 50%)
Outside vapour pressure = 600 Pa (air temperature = 0 °C; R.H. = 100%)

Vapour pressure difference, P = 1340 – 600 = 740 Pa

Material or layer	Thickness, d (m)	Vapour resistivity, r_v (MN s/g m)	Vapour resistance, R_v (MN s/g)	Vapour pressure drop across layer, P_{drop} (Pa)	Vapour pressure at boundary (Pa)	Dew point (°C)
			$r_v \times d$	$\dfrac{R_v}{R_{v(total)}} \times P$		
Inside surface	–	–	–	–	1340	11.5
Plaster	0.012	60	0.72	$\dfrac{0.72}{11.37} \times 740 = 46.9$	1340 – 46.9 = 1293.1	10.9
Aer. Conc. blocks	0.100	50	5.0	$\dfrac{5.0}{11.37} \times 740 = 325.4$	1293.1 – 325.4 = 967.7	6.7
Mineral wool	0.100	5	0.5	$\dfrac{0.5}{11.37} \times 740 = 32.5$	967.7 – 32.5 = 935.2	6.1
Brickwork	0.103	50	5.15	$\dfrac{5.15}{11.37} \times 740 = 335.2$	935.2 – 335.2 = 600	0
Outside surface	–	–	–		600	0
			$R_{v(total)} = 11.37$			

The dew points are plotted on the cross-section of the wall and joined by straight lines. If the structural temperature falls below the dew point, there is a risk of interstitial condensation. This is shown in Figure 8.14. In this example, the structural temperature falls below the dew point; hence, there is a risk of interstitial condensation. The shaded portion, where the interstitial condensation could occur, exists in the whole length and height of the wall.

Example 8.9 Repeat Example 8.8 if the plaster is covered with vinyl wallpaper (vapour resistance = 10 MN s/g)

Solution:
The first part of the question remains unchanged.
   In the second part, the dew points will change due to the wallpaper, as shown below:

Material or layer	Thickness, d (m)	Vapour resistivity, r_v	Vapour resistance	Vapour pressure drop across layer, P_{drop}	Vapour pressure at boundary (Pa)	Dew-point (°C)
Inside surface	–	–	–	–	1340	11.5
Vinyl wallpaper	–	–	10	$\dfrac{10.0}{21.37} \times 740 = 346.3$	1340 – 346.3 = 993.7	7.1
Plaster	0.012	60	0.72	$\dfrac{0.72}{21.37} \times 740 = 24.9$	993.7 – 24.9 = 968.8	7.0
Aer. Conc. blocks	0.100	50	5.0	$\dfrac{5.0}{21.37} \times 740 = 173.1$	968.8 – 173.1 = 795.7	4.0

Mineral wool	0.100	5	0.5	$\dfrac{0.5}{21.37} \times 740 = 17.3$	795.7 − 17.3 = 778.4	3.7
Brickwork	0.103	50	5.15	$\dfrac{5.15}{21.37} \times 740 = 178.4$	778.4 − 178.4 = 600	0
Outside surface	–	–	–		600	0
			$R_{v(total)} =$ 21.37			

Figure 8.15

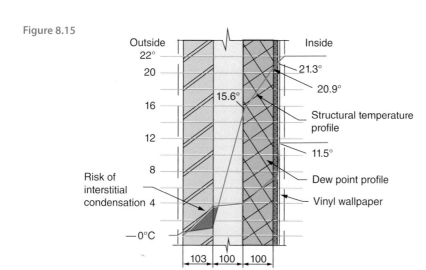

The dew points are plotted on the cross-section of the wall and joined by straight lines, as shown in Figure 8.15. Although not significant, the vinyl wallpaper does reduce the effect of interstitial condensation.

Prevention of Interstitial Condensation

Interstitial condensation can be avoided by keeping the humidity as low as possible. The minimal relative humidity of air required for a comfortable condition is 40%. This can be achieved by keeping a strict control on the sources of moisture so that the amount of water vapour passing through a wall or roof can be minimised. The flow of water vapour can be minimised by providing materials with low permeability, i.e. vapour barriers. These are of two types:

1) Membrane type
   Examples of this type are aluminium foil-backed plasterboard, polythene sheet, asphalt-impregnated paper etc.
2) Paint or liquid type
   Examples of this type are aluminium paint and gloss paint.

The vapour barrier is always applied to the warm side of the element.

Exercise 8.1

1   Find the U-values of a triple-glazed window with two 20 mm wide airspaces.
Given:
Thickness of glass = 4 mm
Thermal conductivity of glass = 1.02 W/m K
Thermal resistance of surfaces: 0.12 (inside); 0.06 (outside)
Thermal resistance of 20 mm wide airspace = 0.18 m^2 K/W

2   **A**   Find the U-value of a 215 mm thick brick wall with a 12 mm layer of dense plaster.
     Thermal conductivities: bricks = 0.84; plaster = 0.46 W/m K
     Use the values of thermal resistance given in Question 1.
   B   The U-value of the wall in Question 2(a) needs to be improved to 0.35 by using sheets of expanded polystyrene (EPS) and 9 mm thick plasterboard. Find the minimum thickness of EPS required to achieve the new U-value. Thermal conductivities: EPS = 0.033; plasterboard = 0.16 W/m K.

3   **A**   Calculate the U-value of a cavity wall with a half-brick-thick (103 mm) outer leaf, a 50 mm wide cavity, a 100 mm thick aerated concrete block inner leaf and a 12 mm thick layer of lightweight plaster.
     Thermal conductivities: aerated concrete blocks = 0.11; bricks = 0.84; lightweight plaster = 0.16 W/m K.
     Thermal resistances of surfaces: 0.12 (inside); 0.06 (outside)
     Thermal resistance of airspace = 0.18 m^2 K/W
   B   Find the U-value of the wall if the cavity is completely filled with mineral wool insulation. The thermal conductivity of mineral wool is 0.04 W/m K.

4   Calculate the total rate of heat loss from the building shown in Figure 8.7. A temperature of 22 °C is to be maintained inside the building when the outside air temperature is 0 °C. Allow 1 air change per hour; the volumetric specific heat capacity of air is 1212 J/m^3 K. The U-values (W/m^2 K) are:
walls = 0.35; floor = 0.25; roof = 0.25; doors = 2.2; windows = 1.9.
The ground temperature is 5 °C.

5   The cavity wall of a dwelling house is constructed of bricks, aerated concrete blocks, lightweight plaster and mineral wool. The inside air is at 21 °C and 60% relative humidity. The outside air is at 0 °C and 100% relative humidity. The properties of the materials and their thicknesses are:

Material	Thickness (d) mm	λ – value W/mK	Vapour resistivity (r_v) MN s/q m
Plaster	12	0.16	60
Aer. Conc. blocks	100	0.11	50
Mineral wool	50	0.04	5
Bricks	103	0.84	50

The cavity is 50 mm wide and filled with insulation. Determine if there is a risk of interstitial condensation.

References/Further Reading

1 CIBSE Guide A (2006). *Environmental Design*. London: Chartered Institution of Building Services Engineers.
2 CIBSE Guide B (2005). *Heating, ventilating, air conditioning and refrigeration*. London: Chartered Institution of Building Services Engineers.
3 Quinn Radiators. Website: www.quinn-radiators.com.
4 Tecpel Co. Ltd., Taiwan. Website: www.tecpel.net.
5 The Building Regulations (2000). *Approved Document L1A – Conservation of Fuel and Power in New Dwellings*. London: Department of Communities and Local Government.

9

Forces and Structures 1

LEARNING OUTCOMES

1) Identify different types of forces (loading) that act on a structure.
2) Explain stress and strain in construction materials.
3) Define elasticity and explain the stress–strain relationship for ductile and brittle materials.
4) Solve problems based on stress, strain, elasticity and factor of safety.

9.1   Introduction

Buildings and civil engineering structures like roads, bridges, dams and tunnels are acted upon by forces which can basically be grouped into two categories: dead loading and imposed loading. These forces, if excessive, may be responsible for the failure of a structure. A building or a civil engineering structure is built to be strong, stable and durable to resist these forces. To achieve these requirements in an economical manner, it is necessary to use materials of appropriate quality and use a mathematical approach for preparing the structural design, which involves the determination of sizes and other attributes of the structural elements. However, before preparing the structural design, it is necessary to analyse the forces acting on a structure. This process is called structural analysis and involves calculating the resultant effect of all the forces which can later be used in the design calculations. For example, the structural analysis of a roof truss involves the calculation of forces in its members due to a combined effect of wind, snow and other loads acting on it. The structural analysis is utilised to determine the sizes of the truss members.

Engineers also deal with the forces acting on retaining walls, multi-storey buildings, bridges, dams and other structures. Each structure is exposed to different types of forces and, therefore, a different approach is required to analyse the forces in each case. A detailed description of the nature of forces and their effects is given in the following sections.

9.2   Force

All structures are acted upon by a variety of forces. Some forces are small in magnitude, causing no significant effect, whereas others can be large and must be considered in the design process. If a stationary body is acted upon by a force, then, depending on the magnitude of

Construction Science and Materials, Second Edition. Surinder Singh Virdi.
© 2017 John Wiley & Sons Ltd. Published 2017 by John Wiley & Sons Ltd.
Companion website: www.wiley.com/go/virdiconstructionscience2e

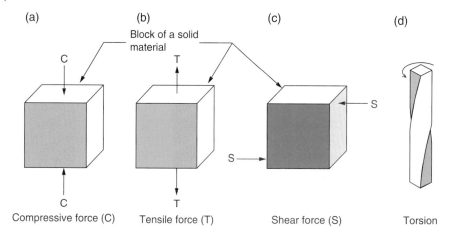

Figure 9.1 Types of force.

the force, it may or may not move. Similarly, a force applied to a moving body may do one or more of the following:

- Change its speed;
- Change its direction;
- Bring it to rest.

Hence, **force** may be defined as that which changes, or tends to change, a body's state of rest or uniform motion in a straight line. A force can be represented in terms of both magnitude and direction. The basic unit of force in the SI system is a newton (N), which can be defined as the force required to produce an acceleration of 1 metre/s^2 in a mass of 1 kg.

The direction of a force can be represented by an arrowhead, for example, a downward force may be represented as ↓ and an upward force as ↑.

Building elements are subjected to the following forces:

1) Push or **compressive force**, which causes shortening of a material.
2) Pull or **tensile force**, which causes stretching of a material.
3) **Shear**, which causes part of the material to move horizontally or vertically against the rest.
4) **Torsion**, which tends to twist a structural member. Some building elements may also be subjected to torsion, for example, edge beams in steel-framed or concrete-framed structures.

These are illustrated in Figure 9.1.

9.2.1 Internal and External Forces

Consider two blocks of a material, as shown in Figure 9.2, one acted upon by a compressive force and the other by a tensile force. These forces are called external forces. The blocks will resist the deforming action of external forces, as internal forces are set up within the material and act in an opposite direction.

9.3 Bending

The effects caused by direct compressive and tensile forces, as shown in Figure 9.1, can also occur in elements subjected to bending. Figure 9.3 illustrates that in a beam, **deflection** (bending) takes place when forces act on it. It is quite evident from the deflected shape (Figure 9.3(b)) that the

Figure 9.2 Internal and external forces.

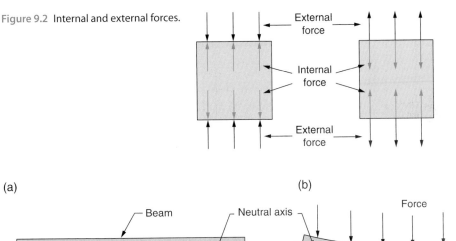

(a)

(b)

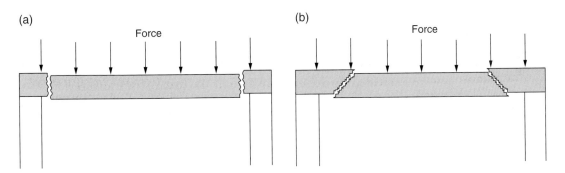

Figure 9.3 Deflection of a simply supported beam.

Figure 9.4 Failure of beams in shear.

upper portion of the beam is shortened or compressed and the lower portion stretched. In other words, we can say that the upper part is in compression, like the material shown in Figure 9.1(a), and the lower part is in tension, like the material shown in Figure 9.1(b). The centre line of the beam remains unchanged in length and that is why it is called the **neutral axis**. It is also evident from Figure 9.3 that compression occurs in the concave part of the beam and tension in the convex part.

The beam is also subjected to **shear**, which is due to the opposite directions of the reaction and the applied force. The maximum shear force to which the beam is subjected is near the supports (see also Chapter 10). In ideal situations, shear failure is assumed to take place at a support (Figure 9.4(a)), but actually it depends on the material of the beam. For example, concrete is assumed to shear at 45°, as shown in Figure 9.4(b).

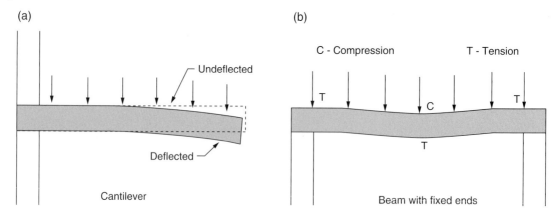

Figure 9.5 Deflection in cantilevers and beams with fixed ends.

The beam shown in Figure 9.3 is a **simply supported** single-span beam, which means that the beam and its supports are not rigidly connected. A beam with one support is called a **cantilever**, which is shown in Figure 9.5(a). The force acting on a cantilever causes tension in the upper half and compression in the lower half.

In reinforced concrete (RC) framed buildings, the beams are considered to have **fixed ends**, i.e. the beams and the columns supporting them are constructed at the same time. The beam ends are not free to move when loading is applied as there are strong and rigid connections between the beams and the columns.

Figure 9.5 shows the deflected shapes of a cantilever and of a beam with fixed ends.

9.3.1   Deflection

Some amount of deflection always occurs in beams and joists when a force (called a load in structural design) is applied on them. The amount of deflection depends on several factors, including:

1) The magnitude of the load.
2) The size of the beam/joist.
3) The properties of the material concerned.

The size of a beam may be adequate to save it from failure, but still a large deflection may be produced. This causes a practical difficulty in securing the finishes to the beams or joists.

The values of actual deflection (δ) of a simply supported beam subjected to a point load or to a uniformly distributed load are:

1) For a point load (central), $\delta = \dfrac{WI^3}{48EI}$
2) For a uniformly distributed load, $\delta = \dfrac{5WI^3}{384EI}$

where E = the modulus of elasticity of the material
       I = the moment of inertia of the beam

The actual deflection of a beam should not be more than 1/360 of the span.

9.4   Types of Loading

A building is acted upon by many forces, which can be divided into two categories:

1) Dead load, imposed load, wind load and loading from other effects.
2) Point load, uniformly distributed load and triangular load.

9.4.1   Dead Load

The dead load of an element or a component is its self-weight, which depends on the density of the materials used and the dimensions of the component. Dead load is of a permanent nature, i.e. it always acts on a structure. For a two-storey dwelling house, the total dead load will consist of:

1) The self-weight of the roof structure due to trussed rafters, roof tiles, battens and ceiling finish.
2) The self-weight of the first-floor construction.
3) The self-weight of the walls, from the top to foundation level.
4) The self-weight of the foundation.

   The calculation of the dead load of an element is very important as it forms a significant part of the design load.

9.4.2   Imposed Load

Any load that is of a temporary nature is called imposed load (or live load). Many types of forces can be considered under this category, for example:

1) Loading from people, furniture, goods etc. in dwelling houses, schools, colleges, hotels, warehouses etc.
2) Loading from people, vehicles and trains etc. on bridges.
3) Snow load: this affects the roof structure of a building.

   Wind load is of a temporary nature as well, but, due to its complexity, it is dealt with separately.

9.4.3   Wind Load

Wind exerts force on buildings, bridges and other objects that stand in its way to obstruct the flow. The force depends on the design of the building, the wind speed and the geographical location. The roof tiles of dwelling houses are especially vulnerable to damage by wind forces.

9.4.4   Loading from other Effects

There are some sources that can form a major part of the total load. Typical examples are: earth pressure in retaining walls, liquid pressure in liquid-retaining structures and seismic load which can cause severe damage to buildings, depending on the intensity of the earthquake.

   For simple structural designs, the dead load and the imposed load are added to find the design load. Figure 9.6 shows the dead and imposed loading acting on a low-rise building.

9.4.5   Point Load (Figure 9.7(a))

The force acting on a very small area is called a **point load**. For example, a person standing on a roof or on a floor is considered a point load.

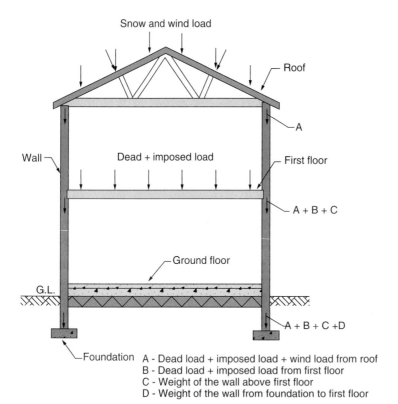

Snow and wind load

Roof

A

Wall

Dead + imposed load

First floor

A + B + C

Ground floor

G.L.

A + B + C + D

Foundation

A - Dead load + imposed load + wind load from roof
B - Dead load + imposed load from first floor
C - Weight of the wall above first floor
D - Weight of the wall from foundation to first floor

Figure 9.6

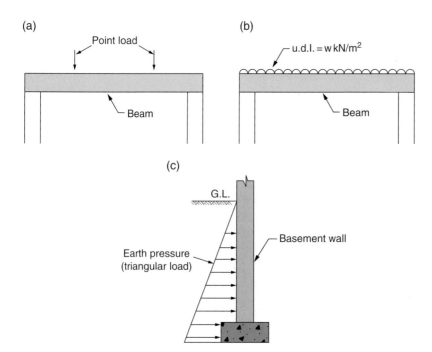

(a)

Point load

Beam

(b)

u.d.l. = w kN/m^2

Beam

(c)

G.L.

Earth pressure
(triangular load)

Basement wall

Figure 9.7 Types of loading.

9.4.6 Uniformly Distributed Load (Figure 9.7(b))

Forces distributed evenly over an area are considered to be a uniformly distributed load (u.d.l.). The resultant force of a u.d.l. acts at the centre of the area on which the load acts.

9.4.7 Triangular Load (Figure 9.7(c))

The force acting on a retaining wall or a basement wall due to the retained soil (called earth pressure) is considered a triangular load. The load is zero at the top and increases as the depth of the retained soil increases. The resultant force acts at a point that is at one-third of the height of the retained soil from the base of the wall.

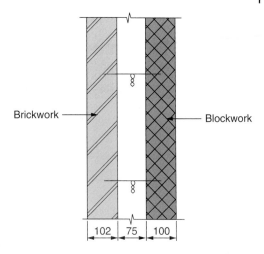

Brickwork

Blockwork

102 75 100

Figure 9.8 Section through a cavity wall.

Example 9.1   Find the dead load of a cavity wall measuring 8.0 m × 5.2 m high. The densities of brickwork and blockwork are 2200 kg/m^3 and 600 kg/m^3, respectively. The cross-section of the wall is shown in Figure 9.8.

Solution:

To find the dead load, we need to find the volume of brickwork and blockwork first, and then use this relationship to find the mass:

$$\text{Density} = \frac{\text{Mass}}{\text{Volume}}$$

$$\text{or, Mass} = \text{Density} \times \text{Volume}$$

$$\text{Volume of brickwork} = \text{length} \times \text{height} \times \text{thickness}$$

$$= 8.0 \times 5.2 \times 0.102 = 4.243\,\text{m}^3$$

$$\text{Volume of blockwork} = 8.0 \times 5.2 \times 0.100 = 4.160\,\text{m}^3$$

$$\text{Mass of brickwork} = 2200 \times 4.243 = 9334.6\,\text{kg}$$

$$\text{Mass of blockwork} = 600 \times 4.160 = 2496\,\text{kg}$$

$$\text{Total mass of the wall} = 11830.6\,\text{kg}$$

$$\text{Weight (or dead load) of the wall} = \text{mass} \times \text{g}$$

$$= 11830.6 \times 9.81$$

$$= 116058.2\,\text{N or }116.058\,\text{kN}$$

Example 9.2   Figure 9.9(a) shows the details of a timber floor. If the densities of timber and plasterboard are 550 kg/m^3 and 950 kg/m^3, respectively, find the dead load of the floor per m^2.

Solution:

Each joist supports the loading from an area enclosed by the dashed lines, shown as the shaded portion in Figure 9.9(b). The width of this portion is 0.4 m, the same as the spacing between the joists.

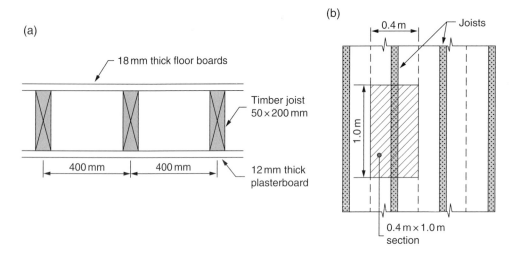

(a)

18 mm thick floor boards

Timber joist
50 × 200 mm

400 mm   400 mm

12 mm thick
plasterboard

(b)

0.4 m   Joists

1.0 m

0.4 m × 1.0 m
section

Figure 9.9 Timber floor.

Consider a 0.4 m × 1.0 m section of the floor; the area of this section is 0.4 m^2

	Volume	Mass
Joist	Volume = length × width × depth = 1.0 × 0.050 × 0.200 = 0.010 m^3	Mass = density × volume = 550 × 0.010 = 5.5 kg
Floor boards	Volume = 1.0 × 0.4 × 0.018 = 0.0072 m^3	Mass = 550 × 0.0072 = 3.96 kg
Plasterboard	Volume = 1.0 × 0.4 × 0.012 = 0.0048 m^3	Mass = 950 × 0.0048 = 4.56 kg

Total mass of floor for an area of 0.4 m^2 = 5.5 + 3.96 + 4.56 = 14.02 kg

Mass of floor per m^2 = 14.02 × 1/0.4 = 35.05 kg

Dead load (or weight) per m^2 = 35.05 × 9.81

= 343.84 N or 0.344 kN

Note: If the room dimensions are known, then determining the number of joists and hence calculating their mass and weight per m^2 will be more accurate.

9.5 Stress and Strain

After finding the total force acting on a structure, it is necessary to find the stress and strain and check whether they are within permissible limits.

9.5.1 Stress

Stress is defined as the force per unit area. In many engineering problems, the effect of force acting on an area can be determined only by calculating the stress produced.

$$\text{Stress} = \frac{\text{Force}}{\text{Area}}$$

The unit of force is the newton (N) or the kilonewton (kN), and the units of area are mm^2 or m^2. By definition, stress is the force (N or kN) per unit area (mm^2 or m^2), therefore, depending on which of these is being used, the units of stress are: N/mm^2, N/m^2, kN/mm^2 or kN/m^2.

Depending on the type of force, stress can be tensile, compressive or shear.

Example 9.3   A column measuring 0.4 m × 0.4 m × 4.0 m high supports a load of 500 kN. Find the stress at the top of the column in kN/m^2 and N/mm^2.

Solution:

Cross-sectional area of the column $= 0.4 \times 0.4 = 0.16 \, m^2$

or, $400 \, mm \times 400 \, mm = 160000 \, mm^2$

$$\text{Stress} = \frac{500}{0.16} = 3125 \text{ kN/m}^2$$

$$= \frac{500 \times 1000 \text{ N}}{160000} = 3.125 \text{ N/mm}^2$$

9.5.2 Strain

A building material deforms when a significant amount of force is applied to it. A tensile force causes elongation of the material and a compressive force causes shortening. As the volume of a material remains constant, the change in its length is accompanied by a change in cross-sectional area.

Strain (also called linear strain) is defined as the ratio of change in length of a material to its original length.

$$\text{Strain}, \varepsilon^* = \frac{\text{Change in length (mm)}}{\text{Original length (mm)}}$$

$^*\varepsilon$ is the Greek letter called epsilon

Strain is specified as a number or percentage, as the change in length and the original length have the same units.

Some materials, like soils, do have a change in volume when forces act on them. The ratio of change in volume to the original volume is known as the **volumetric strain**.

$$\text{Volumetric strain}, \varepsilon_v = \frac{\text{Change in volume}}{\text{Original volume}}$$

Example 9.4   A specimen of a metal, 75 mm long, was stretched with a force of 10 kN. The strain produced in the material was 10%. Find the increase in the length of the specimen.

Solution:

$$\text{Strain}\,(\%) = \frac{\text{Change in length}}{\text{Original length}} \times 100$$

$$10 = \frac{\text{Change in length}}{75} \times 100$$

After transposition, change in length $= \dfrac{10 \times 75}{100} = 7.5\,\text{mm}$

Example 9.5   A specimen of steel measuring 150 mm in length was tested in compression under a load of 20 kN. The final length was 145 mm. Find the strain produced.

Solution:

$$\text{Strain} = \frac{\text{Change in length}}{\text{Original length}}$$

$$= \frac{150 - 145}{150} = 0.033$$

As a percentage, strain $= 0.033 \times 100 = 3.33\%$

9.6   Elasticity

Elasticity can be defined as that property of materials by virtue of which they regain their original shape and size when the deforming forces are removed.

During the seventeenth century, Robert Hooke conducted experiments on metal wires and other types of samples and found that within the elastic range of a material, the change in length was proportional to the force applied. This is known as Hooke's law and can be proven by performing tests on springs, metal wires, strips or solid cylindrical samples.

9.6.1   Experiment 1: Proof of Hooke's Law

A spring is suspended from a fixed point so that only the lower end moves on applying a force. The spring is stretched by placing a mass of 0.250 kg at the lower end and its extension noted (see Figure 9.10). This process is repeated with a range of masses, for example, 0.500 kg, 0.750 kg etc., and the corresponding extensions of the spring are recorded. The mass is converted into weight (force) by multiplying the mass in kilograms by 9.81 (10 is used in Table 9.1 for simplification). A graph is then plotted of force against extension, as shown in Figure 9.11.

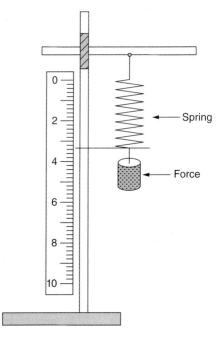

Figure 9.10 Proof of Hooke's law by loading a spring.

A line is drawn through all points and it is observed to be a straight line between the origin and point L. This shows that the extension produced in the spring is proportional to the force applied. The force that corresponds to point L is 12.5 kN. If the force is removed at any stage between the start and 12.5 kN, the spring will return to its original length. The spring behaves like an elastic material up to this point; hence, it is called the **elastic limit**. If the spring is loaded beyond this point, the extension is no longer proportional to the force; the behaviour of the spring is called **plastic behaviour**. If the force is removed during the plastic stage, the spring does not return to its original length.

Table 9.1

Force (N)	Extension (mm)
2.5	5.0
5.0	10.0
7.5	15.0
10.0	20.0
12.5	25.0
15.0	31.0
17.5	42.0

9.6.2 Experiment 2: Proof of Hooke's Law

Hooke's law can also be proven by performing a tensile test on specially prepared metal and non-metal samples. The sample is secured in a piece of apparatus known as a Hounsfield Tensometer and a tensile force is applied by turning the handle. The values of the force and the extension produced in the sample are taken at regular intervals until the sample snaps, and the results of the test are shown either as load versus extension or as stress versus strain. Figure 9.12 illustrates the results of such a test. The graph shows that, as the tensile force is applied, elongation of the sample occurs. The elongation is directly proportional to the force applied, as shown by the straight-line portion, and the point where this behaviour ceases is called the **limit of proportionality (P)**. The material obeys Hooke's law, i.e. it behaves as an elastic material. The ratio of stress and strain within the elastic range is known as **Young's modulus** or the **modulus of elasticity** and is denoted E.

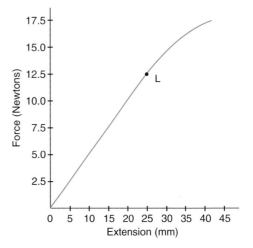

Figure 9.11

$$\text{Young's modulus, E} = \frac{\text{Stress}}{\text{Strain}}$$

As strain is expressed as a number, the unit of Young's modulus is kN/m^2 or kN/mm^2. After the limit of proportionality, the plot is no longer a straight line, but the material still behaves as an elastic material up to a point known as the **elastic limit (E)**. If the loading is removed, the sample will regain its shape and size. The elastic behaviour of the material ceases at the elastic limit and, on further loading, the sample shows plastic behaviour whereby the extension produced is not proportional to the loading. There is another effect produced by plastic behaviour: the sample does not revert to its original shape

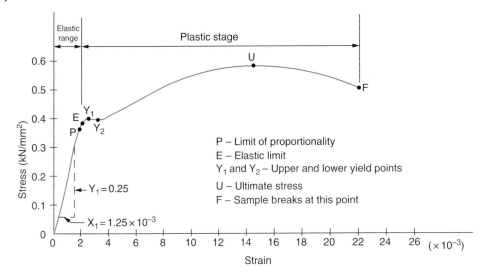

Figure 9.12 Stress–strain relationship for mild steel.

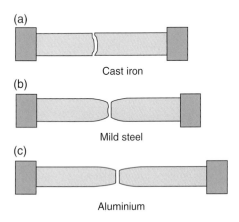

Figure 9.13 Shape of samples after failure.

and size when the load is removed. This permanent increase in the length of the sample is known as a permanent set.

On further increase in load, a point is reached where extension of the sample starts to increase without increment in the load. This point is known as the **upper yield point** (Y_1). At this stage the material will flow (deform) even if the load is reduced slightly. After reaching the **lower yield point** (Y_2), the material regains its resistance to the deforming force. To cause further deformation, the tensile force must be increased. The extension increases at a higher rate due to further increment in the force.

The highest point on the graph corresponds to the maximum load applied and is called the **ultimate stress (U)**. The length of the specimen starts to increase considerably now, with a reduction in the diameter of the sample at a random point. The sample eventually snaps (point F) and the corresponding stress is known as the **breaking stress**. Figure 9.13 shows test specimens of mild steel, aluminium and cast iron after failure. Cast iron, being a brittle material, does not show any elongation, but ductile materials like mild steel, aluminium and copper elongate due to the application of tensile force.

Some typical values of Young's modulus are:

Steel : $210 \, \text{kN/mm}^2$

Aluminium : $70 \, \text{kN/mm}^2$

Copper : $130 \, \text{kN/mm}^2$

Timber : $10 \, \text{kN/mm}^2 \, (\text{variable})$

Example 9.6   A metal rod of 10 mm diameter had an initial length of 1.0 m. After a tensile force of 12 kN was applied to the rod, its length increased to 1.002 m. Calculate Young's modulus of the metal.

Solution:

Cross-sectional area of the rod $= \pi r^2$, where r is the radius

$$= \pi (5)^2 = 78.54 \, \text{mm}^2 \left(\text{radius} = 10 \div 2 = 5 \, \text{mm} \right)$$

$$\text{Stress} = \frac{\text{Force}}{\text{Area}} = \frac{12}{78.54} = 0.1528 \, \text{kN/mm}^2$$

$$\text{Strain} = \frac{\text{Change in length}}{\text{Original length}} = \frac{1.002 - 1.0}{1.0} = 0.002$$

$$\text{Young's modulus} \left(E \right) = \frac{\text{Stress}}{\text{Strain}}$$

$$= \frac{0.1528}{0.002} = 76.4 \, \text{kN/mm}^2$$

Example 9.7   Calculate the value of Young's modulus from the stress–strain graph of steel shown in Figure 9.12.

Solution:
A right-angled triangle of any size is drawn on the straight-line portion of the graph, as shown in Figure 9.12. The length of the vertical side (Y_1) and horizontal side (X_1) are determined and the following formula is used to calculate Young's modulus (E):

$$E = \frac{\text{Stress}}{\text{Strain}} = \frac{Y_1}{X_1}$$

$$= \frac{0.25}{1.25 \times 10^{-3}}$$

$$= 200 \, \text{kN/mm}^2$$

9.6.3   Factor of Safety

Figure 9.12 shows that the maximum tensile stress resisted by a sample is known as the **ultimate stress** (or **ultimate tensile strength**), and is calculated as:

$$\text{Ultimate stress} = \frac{\text{Maximum force carried}}{\text{Original cross-sectional area of sample}}$$

In structural design it is important not to stress any material beyond its elastic limit, otherwise permanent deformation will occur. To ensure that the stress in a material does not exceed its elastic limit, the design calculations are based on a maximum allowable working stress:

$$\text{Maximum allowable working stress} = \frac{\text{Ultimate tensile strength}}{\text{Factor of safety}}$$

Example 9.8   A mild steel tie bar is to withstand a maximum tensile force of 100 kN. The ultimate tensile stress is 540000 kN/m^2 and the factor of safety is to be 3. If Young's modulus of steel is 210 kN/mm^2, calculate:

a) The diameter of the bar.

b) The extension of a 2.0 m long bar when subjected to a force of 100 kN.

Solution:

a) Maximum allowable working stress $= \dfrac{\text{Ultimate tensile strength}}{\text{Factor of safety}}$

$$= \frac{540000}{3} = 180000 \text{ kN/m}^2$$

Maximum allowable working stress $= \dfrac{\text{Maximum tensile force}}{\text{Cross-sectional area of tie bar}}$

$$180000 = \frac{100}{\pi r^2} \left(r = \text{radius} \right)$$

$$r^2 = \frac{100}{\pi \times 180000} = 0.000177$$

$$r = 0.0133 \text{ m}$$

Diameter $= 2 \times r = 2 \times 0.0133 = 0.0266$ m or 26.6 mm

b) Strain $= \dfrac{\text{Change in length}}{\text{Original length}} = \dfrac{\Delta l}{2000} = 0.0005 \Delta l \left(L = 2 \text{ m or 2000 mm} \right)$

Stress $= \dfrac{\text{Force}}{\text{Area}} = \dfrac{100}{\pi \times (0.0133)^2} = 179947.93$ kN/m^2 or 0.18 kN/mm^2

$$E = \frac{\text{Stress}}{\text{Strain}}$$

$$210 = \frac{0.18}{0.0005 \Delta l}$$

$$\Delta l = \frac{0.18}{0.0005 \times 210} = 1.71 \text{ mm}$$

Exercise 9.1

1   A sample of steel is subjected to a compressive force of 4 kN. Calculate the compressive stress in the sample if its cross-section is 10 mm × 10 mm.

2   A 3.0 m long wire is suspended vertically and a load is applied to the bottom of the wire. Find the strain in the wire if it is stretched by 4 mm.

3   A metal rod of 10 mm diameter had an initial length of 2.0 m. After a tensile force of 15 kN was applied to the rod, its length increased to 2.004 m. Calculate Young's modulus of the metal.

4   In a tensile test on a metal, the values of stress and strain were calculated to be:

Stress (kN/mm^2)	Strain ($\times 10^{-3}$)
0	0
0.05	0.3
0.10	0.6
0.15	0.9
0.24	1.5
0.32	2.0
0.39	2.5

Plot a graph between stress and strain and calculate the value of Young's modulus.

5   A 2.5 m long metal bar measuring 15 mm × 10 mm in cross-section is stretched by a force of 15 kN. Find the extension produced if Young's modulus of the metal is 150 kN/mm^2.

6   A metal tie bar is to withstand a maximum tensile force of 100 kN. The ultimate tensile stress is 0.48 kN/mm^2 and the factor of safety is to be 2.5. If Young's modulus of the metal is 180 kN/mm^2, calculate:
   A The diameter of the bar.
   B The extension of a 2.0 m long bar when subjected to a force of 100 kN.

References/Further Reading

1 Durka, F., Al Nageim, H., Morgan, W. and Williams, D. (2002). *Structural Mechanics*. Harlow: Prentice Hall.
2 Hulse, R. and Cain, J. (2003). *Structural Mechanics*. Basingstoke: Palgrave Macmillan.

10

Forces and Structures 2

LEARNING OUTCOMES

1) Calculate the bending moment and shear force in a beam subjected to point and/or uniformly distributed loading.
2) Draw bending moment and shear force diagrams.
3) Apply the law of triangle of forces to solve problems graphically.
4) Determine, graphically, the nature and magnitude of forces in the members of a roof truss.

10.1   Moment of a Force

It has already been discussed in Section 9.3 that a force, or a system of forces, causes deflection in beams and other elements of a building. It is also obvious that the amount of deflection depends on the magnitude of the force, i.e. a small force produces a small deflection as compared to a large force that produces larger deflection. The deflection of a beam is due to the turning or rotational effect of a force. This turning effect is known as the **moment of a force** or the **bending moment**, and depends on:

1) The magnitude of the force/forces.
2) The perpendicular distance between the force and the point about which the rotation or turning takes place.

The above factors can be combined to determine the moment of a force:

Moment of a force = Force × Distance

Figure 10.1 shows two cantilevers, each acted upon by force F. The force tends to cause the rotation of the cantilevers about their supports.

The moment of force F about point x = F × l

where l is the perpendicular distance between the force and point x

The units of force are the newton (N) and the kilonewton (kN); the units of distance are millimetres (mm) and metres (m). As the moment of a force is the product of the force and the distance, its unit could be any of the following:

newton metres (Nm); newton millimetres (Nmm); kilonewton metres (kNm).

Construction Science and Materials, Second Edition. Surinder Singh Virdi.
© 2017 John Wiley & Sons Ltd. Published 2017 by John Wiley & Sons Ltd.
Companion website: www.wiley.com/go/virdiconstructionscience2e

(a)   (b)

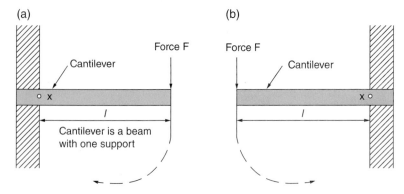

Figure 10.1

In Figure 10.1(a), the effect of the force is to cause clockwise rotation of the cantilever, as shown by the arrow. The moment in this case is known as a clockwise moment. The cantilever shown in Figure 10.1(b) will be subjected to anti-clockwise rotation due to the action of force F, and hence the moment is anti-clockwise.

10.1.1 Sign Convention

In the calculation of moments, it is important to use sign convention to avoid any confusion. Therefore:

> **Clockwise $\left(\text{CW}\right)$ moments are considered to be positive.**
>
> **Anti $-$ clockwise $\left(\text{ACW}\right)$ moments are considered to be negative.**

Example 10.1   Calculate the moment about point x for the cantilever shown in Figure 10.2.

Solution:

Moment about a point = Force × Perpendicular distance

There is only one force (5 kN) acting at a distance of 3 m from point x. Since it tends to cause anti-clockwise rotation of the beam (Figure 10.3), the moment is anti-clockwise.

Moment at $x = 5\,\text{kN} \times 3\,\text{m}$

$\qquad = 15\,\text{kNm}\left(\text{anti-clockwise}\right)$

Figure 10.2

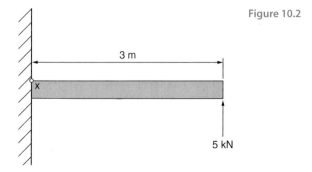

Figure 10.3 Rotational tendency due to 5 kN force.

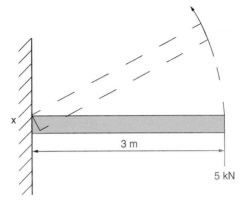

Example 10.2 Calculate the clockwise and anti-clockwise moments about points A and B for the beam shown in Figure 10.4.

Solution:
Three forces are acting on the beam. As the moment about a point = force × distance, we need to determine the moments of all the forces about points B and A.

Moments about point B

1) The 8 N force is acting upwards and would rotate the beam in a clockwise direction if other forces were not present (Figure 10.5). This produces a clockwise moment.

   Clockwise moment about point B, due to the 8 N force
   = force × distance between the 8 N force and point B
   = 8 N × 5 m
   = 40 Nm

2) The 20 N force is acting downwards and would rotate the beam in an anti-clockwise direction if other forces were not present (Figure 10.5). This produces an anti-clockwise moment.

   Anti-clockwise moment about point B, due to the 20 N force
   = force × distance between the 20 N force and point B
   = 20 N × 2 m
   = 40 Nm

Figure 10.4

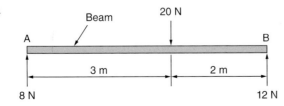

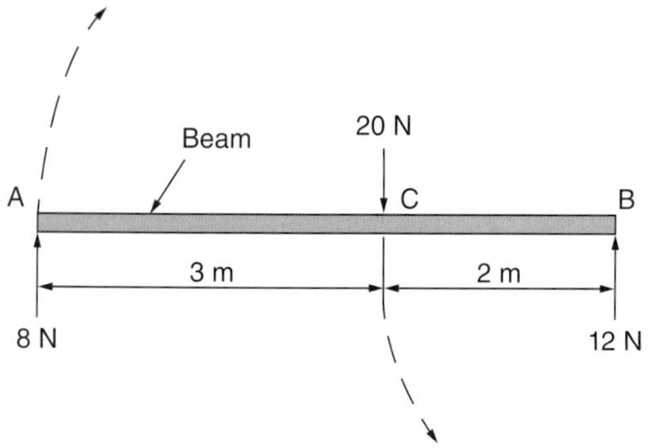

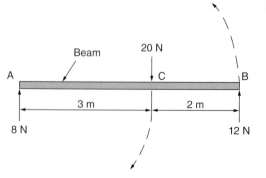

Figure 10.5

3) The 12 N force acts at point B, therefore the distance between the force and point B is zero.

   Moment about point B, due to the 12 N force
   = force × distance between the 12 N force and point B
   = 12 N × 0
   = 0 Nm

Moments about point A

1) The 12 N force is acting upwards and would rotate the beam in an anti-clockwise direction if other forces were not present (Figure 10.6). This produces an anti-clockwise moment.

   Anti-clockwise moment about point A, due to the 12 N force
   = force × distance between the 12 N force and point A
   = 12 N × 5 m
   = 60 Nm

2) The 20 N force is acting downwards and would rotate the beam in a clockwise direction if other forces were not present (Figure 10.6). This produces a clockwise moment.

   Clockwise moment about point A, due to the 20 N force
   = force × distance between the 20 N force and point A
   = 20 N × 3 m
   = 60 Nm

Figure 10.6

3) The 8 N force acts at point A, therefore the distance between the force and point A is zero.

Moment about point A, due to the 8 N force

= force × distance between the 8 N force and point A

= 8 N × 0

= 0 Nm

10.2 Laws of Equilibrium

Building elements like beams, slabs, walls and columns are acted upon by a number of external forces. An element is in equilibrium if it is not disturbed from its state of rest by external forces. In other words, for a building element (or a body) to be in equilibrium, the resultant effect of the forces must be zero. There are three conditions for maintaining the equilibrium of a body:

1) The sum of vertical forces (V) is zero, i.e. $\Sigma V = 0$
   This can also be stated as:
   The sum of downward – acting forces = the sum of upward – acting forces.
2) The sum of horizontal forces (H) is zero, i.e. $\Sigma H = 0$.
3) The sum of moments (M) about any point is zero, i.e. $\Sigma M = 0$
   This can also be stated as:
   The sum of clockwise moments = the sum of anti – clockwise moments.

These conditions will be used in the next section to determine the beam reactions and bending moments.

10.3 Analysis of Beams

In this section we will focus on the analysis of simply supported beams and cantilevers.

10.3.1 Beam Reactions

Newton's third law of motion states that:

To every action there is an equal and opposite reaction.

When a concrete block is placed on the ground, its weight (the action) acts downwards; the ground, due to its strength, offers an upward-acting reaction to counteract the weight of the block. The block will be stable if the reaction equals the weight of the block. If the ground is very loose, the block may sink into it, in which case the action is greater than the reaction.

In the case of a beam, the action is due to the dead, imposed and other loads acting on it. Since beams are supported on walls or columns, the reactions are provided by these supports. The strength of the supports to provide appropriate reaction is of paramount importance. The calculation of beam reactions is necessary for preparing their structural analysis and design, and they are used to check that the stress in the support material is within the permissible limit.

If the loading is symmetrical, then the reactions must be equal. Figure 10.7(a) shows a simply supported beam, resting on two walls, which is acted upon by a 10 kN force at its centre. For structural

(a)

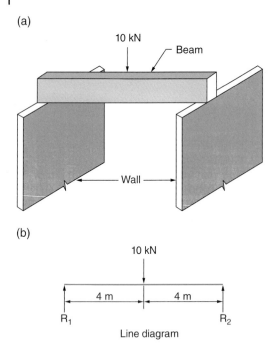

Figure 10.7 Simply supported beam.

(b)

Line diagram

calculations, line diagrams of beams are used; Figure 10.7(b) shows the line diagram of the beam shown in Figure 10.7(a). As the beam is resting on walls, the reactions offered by the walls (R_1 and R_2) must be equal to 10 kN (the action) for the stability of the beam. Or, in other words, the downward-acting force must be equal to the sum of the upward-acting forces.

$$10\,kN = R_1 + R_2$$

As the 10 kN force acts at the centre of the beam, R_1 must be equal to R_2:

$$R_1 = R_2 = 5\,kN$$

Similarly, we can also find reactions if more than one force is acting, as long as the forces are equal and their action points are symmetrical. Figure 10.8 shows a beam symmetrically loaded by two forces of 3 kN each.

$$3\,kN + 3\,kN = R_1 + R_2$$

$$or, 6\,kN = R_1 + R_2$$

$$also, R_1 = R_2$$

$$therefore, R_1 = R_2 = 3\,kN$$

The beams shown in Figure 10.9 are not symmetrically loaded. In these cases, the magnitude of the reactions will not be the same and cannot be calculated as easily as in the previous cases. However, moments of the forces can be taken about the supports to determine the reactions, as explained in Example 10.3.

Figure 10.8

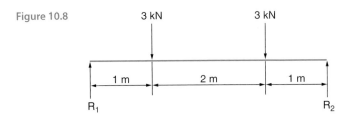

(a)   (b)

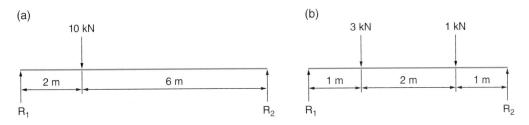

Figure 10.9

Example 10.3   Figure 10.10 shows the line diagram of a beam and the forces acting on it. Calculate reactions R_1 and R_2.

Figure 10.10

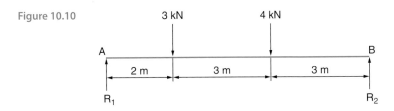

Solution:
As the loading is not symmetrical, R_1 and R_2 will be unequal, but can be determined by taking moments of the forces about B and/or A.

Calculation of reaction R_1
The forces tend to rotate the beam, about point B, in the directions illustrated in Figure 10.11. Reaction R_1 can be calculated by taking moments of all forces about point B and either equating clockwise moments to anti-clockwise moments or taking the algebraic sum and equating it to zero. We'll use the former approach.

   R_1 will cause a clockwise moment, and both the 3 kN and the 4 kN forces will cause anti-clockwise moments.

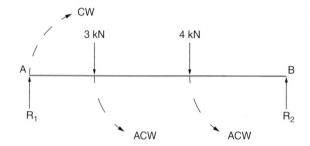

Figure 10.11 Rotational tendency of the forces about B.

$$\text{Clockwise moment} = R_1 \times \text{distance between } R_1 \text{ and point B}$$
$$= R_1 \times 8\,\text{m}$$

Anti-clockwise moment due to the 3 kN force
$$= 3\,\text{kN} \times \text{distance between 3 kN force and point B}$$
$$= 3\,\text{kN} \times 6\,\text{m} = 18\,\text{kNm}$$

Anti-clockwise moment due to the 4 kN force
$$= 4\,\text{kN} \times \text{distance between 4 kN force and point B}$$
$$= 4\,\text{kN} \times 3\,\text{m} = 12\,\text{kNm}$$

$$\text{Total anti-clockwise moment} = 18 + 12 = 30\,\text{kNm}$$

The moment due to R_2 is zero as it acts at point B (see Example 10.2 for explanation).

The clockwise moments must be equal to the anti-clockwise moments to maintain the stability of the beam at point B:

$$R_1 \times 8\,\text{m} = 30\,\text{kNm}$$
$$\text{or,} \quad R_1 = \frac{30\,\text{kNm}}{8\,\text{m}} = 3.75\,\text{kN}$$

Calculation of reaction R_2

Figure 10.12 shows the directions in which the forces tend to rotate the beam about point A. Reaction R_2 will produce an anti-clockwise moment whilst the 3 kN force and the 4 kN force produce clockwise moments.

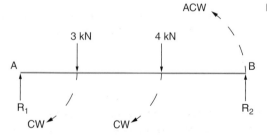

Figure 10.12 Rotational tendency of the forces about A.

Take moments of all forces about point A:

Anti-clockwise moment due to reaction $R_2 = R_2 \times$ distance between R_2 and point A

$$= R_2 \times 8\,m$$

Clockwise moment due to the 4 kN force

$$= 4\,kN \times \text{distance between } 4\,kN \text{ force and point A}$$
$$= 4\,kN \times 5\,m = 20\,kNm$$

Clockwise moment due to the 3 kN force

$$= 3\,kN \times \text{distance between } 3\,kN \text{ force and point A}$$
$$= 3\,kN \times 2\,m = 6\,kNm$$

Total clockwise moment $= 20 + 6 = 26\,kNm$

The moment due to R_1 is zero as it acts at point A (see Example 10.2 for explanation).

The clockwise moments must be equal to the anti-clockwise moments to maintain the stability of the beam at point A:

$$R_2 \times 8\,m = 26\,kNm$$
$$\text{or, } R_2 = \frac{26\,kNm}{8\,m} = 3.25\,kN$$

Check: The sum of reactions R_1 and R_2 must be equal to the total loading of 7 kN

$$R_1 + R_2 = 3.75 + 3.25 = 7\,kN$$

Example 10.4   Calculate reactions R_1 and R_2 for the beam shown in Figure 10.13. A uniformly distributed load (u.d.l.) of 2 kN/m acts on the beam.

Solution:

In this example the loading is symmetrical, which means that the reactions will be equal. The total force (or load) acting on the beam is 2 kN/m $\times$ 6 m = 12 kN. Therefore:

$$R_1 = R_2 = \frac{12}{2} = 6\,kN$$

In practice, loading is usually non-symmetrical; therefore, the method of moments is used to determine the reactions. In order to determine the reactions and to make it easier to calculate the moments, the u.d.l. is converted into a point load.

Total load on the beam $= 12\,kN$

Figure 10.13

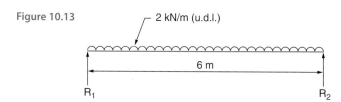

2 kN/m (u.d.l.)

6 m

R_1 R_2

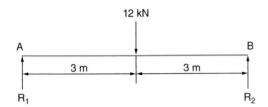

12 kN

A   B

3 m   3 m

R_1 R_2

Figure 10.14 U.d.l. converted into a point load.

The total load (point load) is assumed to act at the centre of the distance over which the u.d.l. acts, as shown in Figure 10.14.

Take moments about B:

Clockwise moment $= R_1 \times 6\,\text{m}$

Anti-clockwise moment $= 12\,\text{kN} \times 3\,\text{m} = 36\,\text{kNm}$

For equilibrium, clockwise moment = anti-clockwise moment

$R_1 \times 6\,\text{m} = 36\,\text{kNm}$

or, $R_1 = \dfrac{36\,\text{kNm}}{6\,\text{m}} = 6\,\text{kN}$

Take moments about A:

Anti-clockwise moment $= R_2 \times 6\,\text{m}$

Clockwise moment $= 12\,\text{kN} \times 3\,\text{m} = 36\,\text{kNm}$

For equilibrium, anti-clockwise moment = clockwise moment

$R_2 \times 6\,\text{m} = 36\,\text{kNm}$

or, $R_2 = \dfrac{36\,\text{kNm}}{6\,\text{m}} = 6\,\text{kN}$

10.3.2   Shear Force (S.F.)

As explained in Chapter 9, a **shear force** causes or tends to cause the horizontal or vertical movement of a part of the material against the rest. The values of shear force and bending moments, due to a set of loads, are used for designing beams, slabs and other building elements. The results of shear force and bending moments are shown in the form of a shear force diagram and a bending moment diagram, respectively.

The shear force at any point on a beam is defined as the algebraic sum of all the forces acting on one (either) side. Figure 10.15 shows the sections of a typical beam where failure due to shear force may occur.

Shear force at point C $=$ algebraic sum of all forces to the left of point C

$= R_1$

Shear force at point D $=$ algebraic sum of all forces to the left of point D

$= R_1 - F$

Figure 10.15 Beam failure due to shear force.

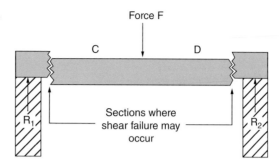

Shear force at point D may also be calculated by taking the algebraic sum of all the forces that act to the right of point D. There is only one force to the right of point D, i.e. R_2. Therefore, shear force at D is also equal to R_2.

Sign Convention
The use of sign convention is important in drawing the shear force diagram without any complexity and confusion. Shear force is considered positive if the forces act 'up on the left and down on the right'. This can also be represented as ↑↓.

Shear force is negative if the forces act down on the left and up on the right (↓↑).

To produce a shear force diagram, a base line is drawn first, and lines are drawn up for upward-acting forces and down for downward-acting forces.

Example 10.5 Draw the shear force diagram for a 6 m long beam carrying a point load of 10 kN at its centre.

Solution:
The beam is shown in Figure 10.16(a). Before producing the shear force diagram, we need to calculate reactions R_1 and R_2. The process has already been explained in Section 10.3.1. Here, $R_1 = R_2 = 5$ kN.

To draw the S.F. diagram, draw up for upward-acting forces and down for downward-acting forces. The process is explained below:

1) Select a suitable scale, for example, 1 kN = 4 mm (see Figure 10.16(b)).
2) Draw base line ab. As a 5 kN force (R_1) acts upwards at A, draw line ac = 20 mm (5 kN = 20 mm) vertically up from the base line.
3) To determine the shear force at point J, which is just to the left of the 10 kN force, take the algebraic sum of all forces acting to the left of point J. There is only one force, i.e. reaction R_1, that acts to the left of point J. Therefore, shear force at J is equal to reaction R_1, i.e. 5 kN.
4) From point c, draw a horizontal line to point d. This shows that the shear force from A to J is the same.
5) The 10 kN force acts downwards at point C. From point d, draw vertical line df to represent the downward-acting 10 kN force. Measure df = 40 mm (1 kN = 4 mm, therefore 10 kN = 40 mm).
6) At point K, which is just to the right of point C, the shear force is the algebraic sum of forces acting to its left, i.e. 5 kN – 10 kN = –5 kN. As there is no force between points K and B, the shear force remains –5 kN. Draw horizontal line fg to represent this.

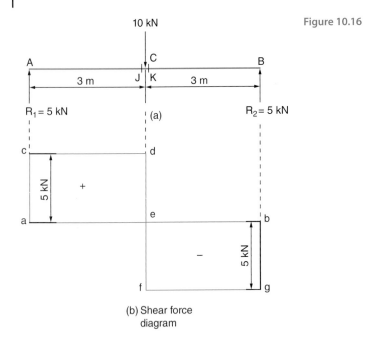

Figure 10.16

(b) Shear force
diagram

7) At point B on the beam, reaction R_2 (5 kN) acts upwards. From point g, draw a vertical line up measuring 20 mm (5 kN = 20 mm) to represent the upward-acting reaction R_2. If this line meets the base line at point b, then the shear force diagram is accurate.

8) Figure 10.16(b) shows the positive and negative portions of the shear force diagram. These have been decided after considering the direction of the forces acting on beam AB.

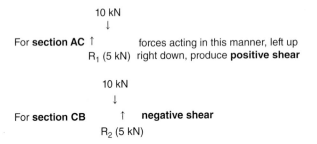

Example 10.6   Draw the shear force diagram for the beam shown in Figure 10.17(a).

Solution:

1) The beam is shown in Figure 10.17(a). Calculate reactions R_1 and R_2 as explained in Example 10.3.

   $R_1 = 3.75$ kN and $R_2 = 3.25$ kN.

2) Use a convenient scale to draw the shear force diagram, say 1 kN = 6.3 mm.
3) Draw base line ab, as shown in Figure 10.17(b). As reaction R_1 acts upwards, draw line ac = 23.6 mm (3.75 kN × 6.3 mm = 23.6 mm). 23.6 mm represents 3.75 kN, the shear force at A.

Figure 10.17

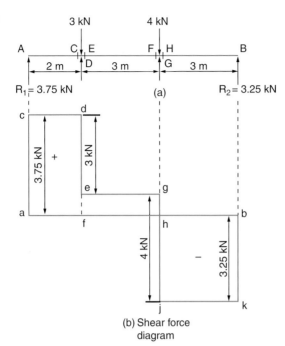

(a)

$R_1 = 3.75$ kN

$R_2 = 3.25$ kN

(b) Shear force
diagram

4) To determine the shear force at any point between A and D, consider point C which is just to the left of point D and take the algebraic sum of all the forces acting between A and C. There is only one force, i.e. 3.75 kN, that is the shear force between A and C.

5) Draw line cd horizontally to show the shear force in this section is 3.75 kN.

6) At point D, there is a downward-acting force of 3 kN. To represent this force, draw down from point d vertical line de measuring 18.9 mm (3 kN × 6.3 mm = 18.9 mm).

7) Shear force at point E, which is just to the right of the 3 kN force, is the algebraic sum of forces acting to the left of E, i.e. 3.75 − 3.0 = 0.75 kN.

8) Shear force at point F = 3.75 − 3.0 = 0.75 kN, which is equal to the S.F. at E. Draw a horizontal line from point e to g showing that the S.F. between E and F is 0.75 kN.

9) A 4 kN force acts downwards at G; draw vertically down from g line gj = 25.2 mm (4 kN × 6.3 mm = 25.2 mm).

10) Shear force at H, which is immediately to the right of point G = 3.75 − 3 − 4 = −3.25 kN. Line hj represents a shear force of −3.25 kN.

11) Shear force at a point immediately to the left of point B = 3.75 − 3 − 4 = −3.25 kN. This is equal to the S.F. at H, therefore draw a horizontal line from j to k representing a shear force of −3.25 kN.

12) Reaction R_2 (3.25 kN) acts upwards at B, therefore draw a 20.5 mm (3.25 kN × 6.3 mm = 20.5 mm) long vertical line from k. If the shear force diagram is accurate, the other end of this line will coincide with point b.

13) The positive and the negative portions of the diagram are decided on the same basis as that used in Example 10.5.

Example 10.7   Draw the shear force diagram for a 6 m long beam carrying a u.d.l. of 2 kN/m.

Solution:

1) The beam with its loading is shown in Figure 10.18(a). Reaction $R_1 = R_2 = 6$ kN.
2) Use a convenient scale, say 1 kN = 3 mm.
3) As a u.d.l. is analogous to closely spaced point loads (Figure 10.18(b)), it is necessary to convert it to a point load for calculating the shear force at a point.
4) Draw horizontal line ab as the base line (Figure 10.18(b)). As reaction R_1 acts upwards, draw vertical line af up from point a to represent it. Measure af = 18 mm to represent R_1 (6 kN × 3 mm = 18 mm) and hence the shear force at A.
5) The S.F. at point C = net vertical force to the left of C (AC = CD = DE = EB = 1.5 m)

$$= R_1 - \text{downward-acting force between A and C}$$
$$= 6\,\text{kN} - (2\,\text{kN/m} \times 1.5\,\text{m})$$
$$= 6\,\text{kN} - 3\,\text{kN} = 3\,\text{kN}$$

(a)   Figure 10.18

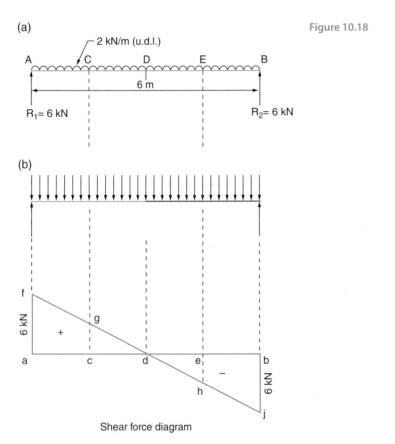

Shear force diagram

6) Mark point g vertically below C to represent the S.F. at C. Distance gc should be 9 mm (3 kN = 9 mm on the shear force diagram).

7) Shear force at D = net vertical force to the left of point D

$$= R_1 - \text{downward-acting force between A and D}$$
$$= 6\,\text{kN} - (2\,\text{kN/m} \times 3\,\text{m})$$
$$= 6\,\text{kN} - 6\,\text{kN} = 0\,\text{kN}$$

Mark point d on the base line, vertically below D.

8) Shear force at E = net vertical force to the left of point E

$$= R_1 - \text{downward-acting force between A and E}$$
$$= 6\,\text{kN} - (2\,\text{kN/m} \times 4.5\,\text{m})$$
$$= 6\,\text{kN} - 9\,\text{kN} = -3\,\text{kN}$$

As the S.F. is negative, this part of the S.F. diagram will be drawn below the base line. Mark point h so that eh = 9 mm (3 kN × 3 mm = 9 mm).

9) Shear force at B = net vertical force to the left of point B

$$= R_1 - \text{downward-acting force between A and B}$$
$$= 6\,\text{kN} - (2\,\text{kN/m} \times 6\,\text{m})$$
$$= 6\,\text{kN} - 12\,\text{kN} = -6\,\text{kN}$$

Mark point j so that bj = 18 mm (6 kN × 3 mm = 18 mm).

10) Join points f, g, d, h and j to produce a straight line.

11) The shear force in the left half of the beam is positive and in the right half, negative. This is illustrated in Figures 10.18(b) and 10.19.

Figure 10.19

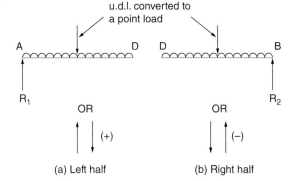

(a) Left half   (b) Right half

10.3.3 Bending Moment (B.M.)

Vertical loading causes, or tends to cause, bending in beams. The amount of bending depends on the magnitude of loading, the span and size of the beam. The bending moment is a measure of the amount of bending at a point on the beam. It is calculated by taking the algebraic sum of moments about the point under consideration. Depending on the type of beam, the forces can cause either

(a)

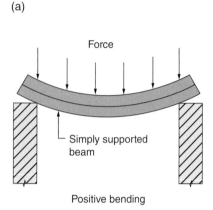

Positive bending

(b)

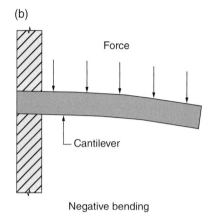

Negative bending

Figure 10.20

sagging or hogging of the beam. Figure 10.20 shows bending in a simply supported beam and a cantilever. In producing the B.M. diagrams:

- A sagging moment is considered positive;
- A hogging moment is considered negative.

Also, as described previously, **clockwise (CW)** moments are taken to be **positive**; **anti-clockwise (ACW)** moments are taken to be **negative**.

The structural design of a beam is based on the maximum bending moment, which can be determined from the B.M. diagram. The purpose of the next three examples is to show the calculation of the bending moment at various points and produce B.M. diagrams.

Example 10.8   Draw the bending moment diagram for the beam shown in Figure 10.21(a).

Solution:
Before the B.M. diagram can be drawn, we need to calculate the bending moment at various points, for example, A, C and B. To determine the B.M. at any point, take moments of all forces to its left and find their algebraic sum.

1) Determine reactions R_1 and R_2, as explained earlier. $R_1 = R_2 = 5$ kN.
2) **B.M. at A:** To calculate the bending moment at point A, we need to take moments of all forces to its left. As there is no force to the left of point A, the B.M. at A is zero. Although reaction R_1 acts at A, the moment produced is zero:

$$\text{Moment about A} \left(M_A \right) \text{due to } R_1 = \text{Force } R_1 \times \text{distance between } R_1 \text{ and A}$$
$$= 5\,\text{kN} \times 0 = 0\,\text{kNm}$$

3) **B.M. at C:** Consider all forces to the left of point C and take the algebraic sum of their moments about C. There is only R_1 that acts on the left of C, and it produces a clockwise moment (positive):

$$\text{Moment about C} \left(M_C \right) \text{due to } R_1 = \text{Force } R_1 \times \text{perpendicular distance between } R_1 \text{and point C}$$
$$= 5\,\text{kN} \times 4\,\text{m} = 20\,\text{kNm}$$

Figure 10.21

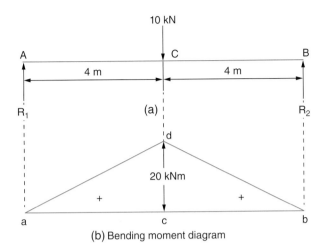

(b) Bending moment diagram

4) **B.M. at B:** Consider all forces to the left of point B. R_1 and 10 kN act to the left of B; they produce clockwise and anti-clockwise moments, respectively.

Clockwise moment about $B = R_1 \times$ perpendicular distance between R_1 and point B

$$= 5\,kN \times 8\,m = 40\,kNm\,(CW\text{ moment is positive})$$

Anti-clockwise moment about $B = 10\,kN \times$ perpendicular distance between 10 kN force and B

$$= 10\,kN \times 4\,m$$

$$= 40\,kNm\,(ACW\text{ moment is considered negative})$$

R_2 will produce a zero moment about B as its distance from B is zero.

B.M. at $B\,(M_B)$ = algebraic sum of moments about B

$$= CW\text{ moment} - ACW\text{ moment}$$

$$= 40 - 40 = 0\,kNm.$$

5) The B.M. diagram, like the S.F. diagram, is drawn below the line diagram of the beam.
Select a suitable scale, say 1 kNm = 1 mm
Bending moment at C = 20 kNm; 20 kNm = 20 mm
Draw base line acb and mark point d so that cd = 20 mm. Complete the B.M. diagram, as shown in Figure 10.21(b). The positive bending moment has been shown above the base line.

Example 10.9 Draw the bending moment diagram for the beam shown in Figure 10.22(a).

Solution:
1) Calculate reactions R_1 and R_2 (refer to Example 10.3)

$$R_1 = 3.75\,kN\text{ and }R_2 = 3.25\,kN$$

2) To determine the bending moment at point A, take moments of forces that act to its left. As there are no such forces, the B.M. at A is zero.

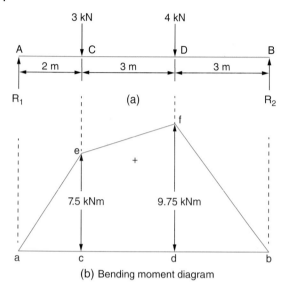

Figure 10.22

(a)

(b) Bending moment diagram

3) **B.M. at C:** Take moments of all forces that act to the left of point C, about point C, and find their algebraic sum. There is only one force (R_1) that will produce a clockwise moment.

$$\text{Clockwise moment due to } R_1 = R_1 \times \text{perpendicular distance between } R_1 \text{ and point C}$$

$$= 3.75\,\text{kN} \times 2\,\text{m} = 7.5\,\text{kNm}$$

The moment due to the 3 kN force is zero as the distance between the force and point C is zero. Therefore, the B.M. at C (M_C) = 7.5 kNm (positive).

4) **B.M. at D:** Find the algebraic sum of moments about point D, due to the forces that act to its left. R_1 and 3 kN act to the left of D and produce clockwise and anti-clockwise moments, respectively:

$$\text{CW moment due to } R_1 = R_1 \times \text{perpendicular distance between } R_1 \text{ and point D}$$

$$= 3.75\,\text{kN} \times 5\,\text{m} = 18.75\,\text{kNm}$$

$$\text{ACW moment due to } 3\,\text{kN} = 3\,\text{kN} \times \text{perpendicular distance between } 3\,\text{kN and point D}$$

$$= 3\,\text{kN} \times 3\,\text{m} = 9\,\text{kNm}$$

The moment due to the 4 kN force that acts at D is zero as the distance between the force and point D is zero.

$$\text{Therefore, the B.M. at } D\left(M_D\right) = \text{algebraic sum of all moments about D}$$

$$= \text{CW moment} - \text{ACW moment}$$

$$= 18.75 - 9.0 = 9.75\,\text{kNm}$$

5) **B.M. at B:** Take moments about B for all forces that act to the left of point B:

CW moment due to $R_1 = R_1 \times$ perpendicular distance between R_1 and point B

$$= 3.75 \, \text{kN} \times 8 \, \text{m} = 30 \, \text{kNm}$$

ACW moment due to 3 kN = 3 kN $\times$ perpendicular distance between 3 kN and point B

$$= 3 \, \text{kN} \times 6 \, \text{m} = 18 \, \text{kNm}$$

ACW moment due to 4 kN = 4 kN $\times$ distance DB

$$= 4 \, \text{kN} \times 3 \, \text{m} = 12 \, \text{kNm}$$

Therefore, the B.M. at B (M_B) = CW moment − ACW moments

$$= 30 - 18 - 12 = 0 \, \text{kNm}$$

6) Select a suitable scale, say 1 kNm = 4 mm

$$7.5 \, \text{kNm} = 7.5 \times 4 = 30 \, \text{mm}$$
$$9.75 \, \text{kNm} = 9.75 \times 4 = 39 \, \text{mm}$$

Draw base line acdb. Mark points e and f so that:

$$ce = 30 \, \text{mm} \, (\text{B.M. at C} = 7.5 \, \text{kNm})$$
$$df = 39 \, \text{mm} \, (\text{B.M. at D} = 9.75 \, \text{kNm})$$

Complete the diagram, as shown in Figure 10.22(b).

Example 10.10   Draw the bending moment diagram for a 6 m long beam carrying a u.d.l. of 2 kN/m.

Solution:
The beam is shown in Figure 10.23. Distances AC = CD = DE = EB.

1) Calculate reactions R_1 and R_2, as explained in Example 10.4

$$R_1 = R_2 = 6 \, \text{kN}$$

2) **B.M. at A:** As explained in Examples 10.8 and 10.9, the bending moment at A is zero as there is no force acting to the left of point A.
3) **B.M. at C:** Take moments about point C for all forces acting to its left, and find their algebraic sum. Before the moments can be taken, the u.d.l. needs to be converted to a point load. The u.d.l. acting to the left of C is equivalent to a point load of:

$$2 \, \text{kN/m} \times 1.5 \, \text{m} = 3 \, \text{kN}$$

Figure 10.23

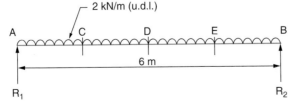

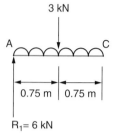

Figure 10.24 Bending moment at C.

3 kN

A ⌒⌒⌒⌒ C

0.75 m | 0.75 m

$R_1 = 6$ kN

The force of 3 kN is supposed to act at the mid-point of AC, as shown in Figure 10.24.

CW moment about $C = R_1 \times$ distance AC

$$= 6\,\text{kN} \times 1.5\,\text{m} = 9.0\,\text{kNm}$$

ACW moment about $C = 3\,\text{kN} \times$ perpendicular distance between 3 kN force and point C

$$= 3\,\text{kN} \times 0.75\,\text{m} = 2.25\,\text{kNm}$$

B.M. at C = algebraic sum of CW and ACW moments

$$= 9.0 - 2.25 = 6.75\,\text{kNm}$$

4) **B.M. at D:** Repeat Step 3 to determine the bending moment at point D. The u.d.l. acting to the left of D (between A and D) is equivalent to a point load of 6 kN (2 kN/m × 3 m = 6 kN). This is also assumed to act at the mid-point of AD (Figure 10.25).

CW moment about $D = R_1 \times$ distance AD

$$= 6\,\text{kN} \times 3\,\text{m} = 18\,\text{kNm}$$

ACW moment about $D = 6\,\text{kN} \times$ perpendicular distance between 6 kN force and point D

$$= 6\,\text{kN} \times 1.5\,\text{m} = 9.0\,\text{kNm}$$

B.M. at D = algebraic sum of CW and ACW moments

$$= 18 - 9.0 = 9.0\,\text{kNm}$$

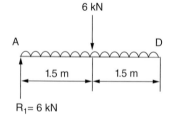

Figure 10.25 Bending moment at D.

6 kN

A ⌒⌒⌒⌒⌒⌒ D

1.5 m | 1.5 m

$R_1 = 6$ kN

5) **B.M. at E:** Repeat the procedure already used in Steps 3 and 4. The u.d.l. to the left of E (between A and E) is equivalent to a point load of 9 kN (2 kN/m × 4.5 m = 9 kN). This acts at the mid-point of AE (Figure 10.26).

Figure 10.26 Bending moment at E.

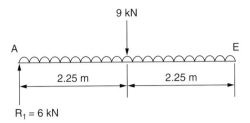

CW moment about $E = R_1 \times$ distance AE

$$= 6\,kN \times 4.5\,m = 27\,kNm$$

ACW moment about $E = 9\,kN \times$ perpendicular distance between 9 kN force and point E

$$= 9\,kN \times 2.25\,m = 20.25\,kNm$$

B.M. at $E =$ algebraic sum of CW and ACW moments

$$= 27 - 20.25 = 6.75\,kNm$$

6) **B.M. at B:** Convert the u.d.l. acting to the left of B into an equivalent point load:

$$= 2\,kN/m \times 6\,m = 12\,kN$$

This is supposed to act at the mid-point of AB, i.e. at D (Figure 10.27).

CW moment about $B = R_1 \times$ distance AB

$$= 6\,kN \times 6\,m = 36\,kNm$$

ACW moment about $B = 12\,kN \times$ perpendicular distance between 12 kN force and point B

$$= 12\,kN \times 3.0\,m = 36\,kNm$$

The moment due to R_2 is zero, as the distance between R_2 and B is zero.

B.M. at $B =$ algebraic sum of CW and ACW moments

$$= 36 - 36 = 0\,kNm$$

The bending moment diagram is shown in Figure 10.28(b).

Note: If a u.d.l. acts on the whole length of the beam, and no other force is involved, then the maximum B.M. (M_{max}) can be calculated from the formula:

$$M_{max} = \frac{Wl^2}{8},$$ where w is the u.d.l. and l is the length of the beam.

In Example 10.10, $M_{max} = \dfrac{2 \times 6^2}{8} = 9\,kNm$

Figure 10.27 Bending moment at B.

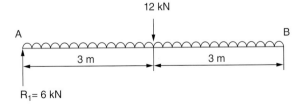

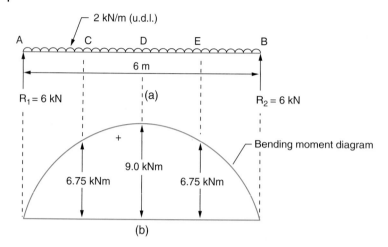

Figure 10.28

10.4 Triangle of Forces

A force can be represented in terms of both magnitude and direction, and hence is a vector quantity. Units of newtons (N), kilonewtons (kN) or Meganewtons (MN) are used to show the magnitude; the direction is shown by using short lines with arrowheads, as explained in Chapter 9.

Forces meeting at a point are known as **concurrent forces**. Forces whose line of action lies in the same plane are called **coplanar forces**. If a body is in equilibrium under the action of three concurrent, coplanar forces, then these forces can be represented by a triangle, with its sides drawn parallel to the direction of the forces. This is known as the law of the **triangle of forces** and can be used to determine the unknown magnitude and/or direction of one or two forces.

If the force system consists of more than three forces, then the law of the polygon of forces can be used, which is similar to the law of the triangle of forces.

If a member is subjected to compression or tension, then the internal resistance of the material acts in the opposite direction, as shown in Figure 10.29. In all problems where we use the triangle of forces or the polygon of forces, we use the direction of internal resistance to show compression or tension.

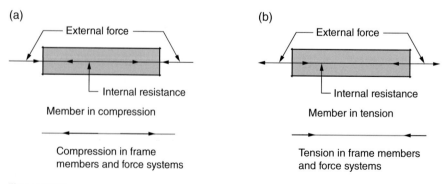

Figure 10.29

Example 10.11   A weight of 20 N is supported by two strings, as shown in Figure 10.30(a). Determine, graphically, the nature and magnitude of forces in the two strings.

Solution:
The 20 N force acts downwards. As the two strings A and B support this force, they must be subjected to upward-acting forces to maintain the equilibrium. Figure 10.30(b) illustrates this by showing details at joint x. For the equilibrium of the system (for the 20 N force to stay in position), the total vertical component of forces in the two strings must be equal to 20 N. The forces in the strings can be determined from the force triangle.
   To draw the force triangle:

1) Select a suitable scale, say $1 N = 2 mm$.
2) Draw line $ab = 40 mm$ ($1 N = 2 mm$, therefore $20 N = 40 mm$) parallel to the 20 N force. The direction of the force in ab should be the same as the direction of the 20 N force.
3) From point b, draw line bc parallel to string A. As we do not know the magnitude of the force in string A, point c cannot be established as yet.
4) Draw a line that passes through point a and is parallel to string B. This will establish point c, as shown in Figure 10.30(c).
5) Triangle abc is the force triangle. Side ab represents the 20 N force, sides bc and ca represent the forces in strings A and B, respectively.
6) Measure bc and ca in millimetres and use the scale of $1 N = 2 mm$ (used to draw the force triangle) to convert these into newtons.

$1 N = 2 mm$ can also be written as $1 mm = 0.5 N$

$bc = 21 mm$, therefore force in string A $= 21 \times 0.5 = 10.5 N$

$ca = 30 mm$, therefore force in string B $= 30 \times 0.5 = 15 N$

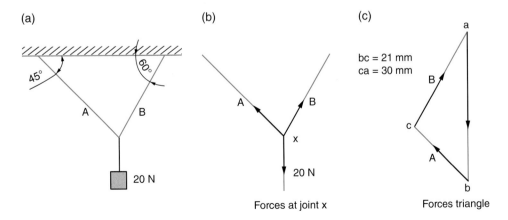

(a)   (b)   (c)

Forces at joint x   Forces triangle

Figure 10.30

Example 10.12   Two rods, A and B, support an object weighing 250 N, as shown in Figure 10.31(a). Determine the nature and magnitude of forces in the two rods.

Solution:
At joint x, the 250 N force acts vertically downwards and is supported by rods A and B. The downward force tends to stretch rod B and hence creates tension. Let us assume that there is tension in rod A as well. The directions of these forces are shown in Figure 10.31(b).
   To draw the force triangle:

1) Only consider the direction of forces at joint x.
2) Select a suitable scale, say 5 N = 1 mm.
3) To represent the 250 N force, draw a vertical line ab = 50 mm (5 N = 1 mm, therefore 250 N = 50 mm). The direction of the force in line ab is downwards, as shown in Figure 10.31(c).

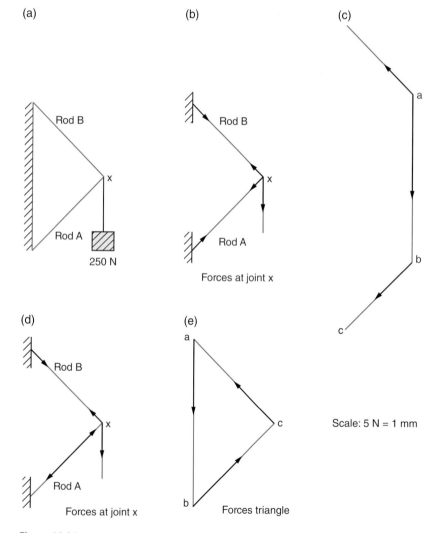

(a)

(b)

(c)

(d)

(e)

Scale: 5 N = 1 mm

Figure 10.31

4) From point b, draw a line parallel to rod A in the downward direction. The downward direction represents the direction of the force in rod A. As the magnitude of the force is to be determined, the length of line bc, which represents the force in the rod, cannot be ascertained as yet.

5) The force in rod B acts upwards, at joint x. To represent this force in the force triangle, a line should be drawn which passes through point a, is parallel to rod B and completes the force triangle. Figure 10.31(c) shows the line that passes through point a and is parallel to rod B. As this does not form a triangle, the assumed direction of the force in rod A is wrong.

6) Change the direction of the force in rod A, as shown in Figure 10.31(d). In order to produce the force triangle, repeat the procedure that has already been explained in Steps 3 to 5.

7) Draw line ab = 50 mm to represent the 250 N force (Figure 10.31(e)). From point b, draw a line parallel to rod A in the upward direction to represent the force in rod A.

8) The next step is to draw a line that passes through point a and is parallel to rod B. This line will also establish point c, as shown in Figure 10.31(e). This procedure results in a triangle, therefore the assumptions regarding the direction of forces in rods A and B are correct.

9) From the force triangle, measure lines bc and ca in mm and, using the scale chosen earlier, convert these to newtons to show the forces in rods A and B.

10) Draw the arrowheads at the other ends of the rods (Figure 10.31(d)). The nature of the forces can be ascertained now.

Force in rod A = 35 mm × 5 N = 175 N COMPRESSION

Force in rod B = 35 mm × 5 N = 175 N TENSION

10.4.1 Bow's Notation

Bow's notation is used to identify forces in frames and force systems. It involves marking the spaces in the space diagram with capital letters, as shown in Figure 10.32. Any force and any member of the frame can be identified by letters, on either side. For example, reaction R_1 is identified as AB as it has spaces A and B on either side. Similarly, the 5 kN force is identified as BC.

It is usual practice to start from the left-hand support and move in a clockwise direction at each joint (node). Although it is not necessary, the first few letters are used exclusively for reactions and external forces.

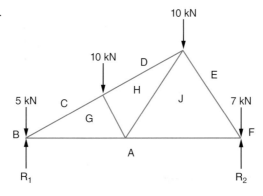

Figure 10.32 Bow's notation.

Example 10.13   For the force system shown in Figure 10.33(a), find the nature and magnitude of the unknown forces.

Solution:

1) Identify the spaces in the space diagram by letters A, B, C and D (Figure 10.33(b)).
2) Use a scale of 1 kN = 8.2 mm. To represent force CD (5 kN), draw line cd parallel to force CD. As 1 kN = 8.2 mm, the length of cd is 41 mm (Figure 10.33(c)).
3) Moving clockwise at joint x, the next member of the force system is DA. As the direction of the force in member DA, at joint x, is downwards, draw line da downwards and parallel to member DA. The direction in which da is drawn should be the same as that of the force in member DA at joint x. The force in member DA is 3 kN, therefore the length of da should be 24.6 mm.
4) The next member we come across is AB. The magnitude and direction of force in this member are not known, therefore draw a line parallel to AB on either side of point a. Point b cannot be marked yet because the force in member AB is not known.
5) The next member of the force system is BC. Again, neither the magnitude nor the direction of the force in this member is known. Draw a line that is parallel to BC and passes through point c. This also establishes point b, as shown in Figure 10.33(c).
6) The directions of forces in the polygon of forces can now be shown. The direction of forces in members of the force system at joint x will be the same as those in the corresponding sides of the polygon of forces.
7) The direction of forces at the other ends of the members should be marked as illustrated in Figure 10.33(b).
8) Measure sides ab and bc of the force polygon in centimetres and convert these to kilonewtons using the chosen scale of 8.2 mm = 1 kN.

Results:

Member	Nature of force	Force (kN)
AB	Compression	6.3
BC	Tension	5.4
DA	Tension	3.0

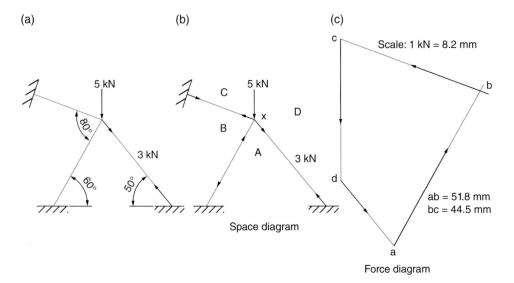

(a)   (b)   (c)

Space diagram

Force diagram

Figure 10.33

10.4.2 Frames and Roof Trusses

Frames and roof trusses contain many members to carry the applied forces in a safe manner. The simplest roof truss consists of three members (Figure 10.34(a)), but a more practical truss, like a trussed rafter, consists of several members (Figures 10.34(b) and 10.34(c)).

The number of joints, also called **nodes**, depends on the number of members of the truss.

In Figure 10.34(a), there are 3 members and 3 joints.

In Figure 10.34(b), there are 11 members and 7 joints.

In Figure 10.34(c), there are 15 members and 9 joints.

Figure 10.34   (a)

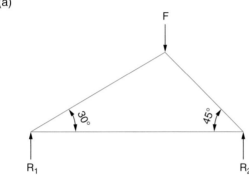

(b)

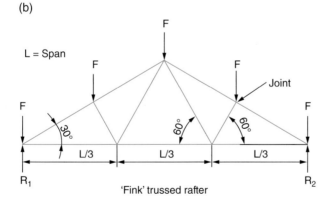

'Fink' trussed rafter

(c)

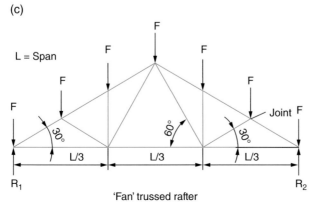

'Fan' trussed rafter

A simple frame may be analysed by considering each joint individually and drawing the force diagrams. For frames with many joints, this would prove to be a lengthy process. To simplify the process, only one force diagram is drawn, as explained in Examples 10.14 and 10.15.

Members of a truss that carry compression are called **struts**. The uppermost members of a truss, called **rafters**, are always in compression. Members in tension are known as **ties**, for instance, the horizontal members at the bottom of a truss.

Example 10.14   Find the nature and the magnitude of the forces in members of the truss shown in Figure 10.35(a).

Solution:

1) Before drawing the force polygon, the magnitudes of reactions R_1 and R_2 should be determined. The truss is not symmetrical; therefore, the law of moments should be used to determine the reactions.

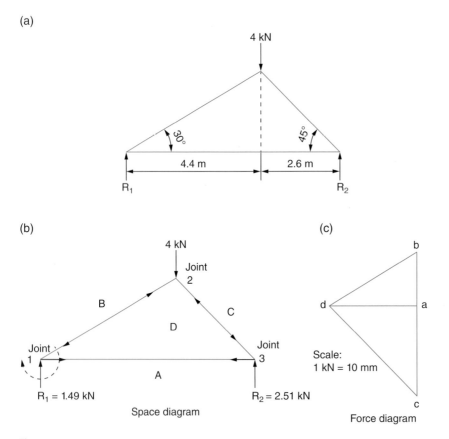

Figure 10.35

2) Take moments of all forces about joint 3.

$$R_1 \times 7.0 = 4 \times 2.6$$

$$\text{Therefore,} \, R_1 = \frac{4 \times 2.6}{7.0} = 1.49 \, \text{kN}$$

$$\text{As} \, R_1 + R_2 = 4.0,$$

$$R_2 = 4.0 - 1.49 = 2.51 \, \text{kN}$$

3) Label the space diagram (Figure 10.35(b)) with capital letters. As explained earlier, members BD and CD are in compression and member DA is in tension. The directions of the forces in the three members are shown in the space diagram.
4) Select a convenient scale, say 1 kN = 10 mm.
5) We have to consider all joints, one by one. Let us start at joint 1 and consider reaction R_1 as its value is known. Draw line ab vertically up to represent force AB (reaction R_1). The magnitude of the force is 1.49 kN, therefore line ab should be 14.9 mm long (Figure 10.35(c)). Line ab is drawn upwards as reaction R_1 (force AB) acts upwards.
6) Moving clockwise at joint 1, as shown by the dotted arrow, the next member of the truss is BD. The direction of force in member BD, at joint 1, is downwards; therefore, draw line bd parallel to BD in the downward direction. Since the magnitude of the force in member BD is not known, point d cannot be established at this stage.
7) Moving clockwise again at joint 1, the next member is DA. As it is horizontal, draw a horizontal line that passes through point a and thus establishes point d on line bd. Lines bd and da will be measured later to determine the magnitude of forces in members BD and DA.
8) Let us move on to joint 2. To represent force BC (a 4 kN force) on the force diagram, draw line bc = 40 mm vertically down, as the force acts downwards.
9) Moving clockwise at joint 2, the next member is CD. The direction of the force in this member is upwards; therefore, draw a line up from point c to represent the force in member CD. Line cd should be parallel to member CD.
10) The next member at joint 2 is DB. Line db, which represents the force in member DB, has already been drawn.
11) At joint 3, we have two members of the truss, AD and DC, and reaction CA. These have already been represented on the force diagram.
12) The force diagram is now complete. Measure sides bd, cd and da, which represent forces in members BD, CD and DA, respectively. Use the scale already selected, i.e. 1 kN = 10 mm, and convert the lengths of the measured sides into kilonewtons. The results are shown below:

Member	Nature of force	Force (kN)
BD	Compression	3.0
CD	Compression	3.6
AD	Tension	2.6

Example 10.15 Find the nature and the magnitude of forces in the members of the trussed rafter shown in Figure 10.36(a).

(a)

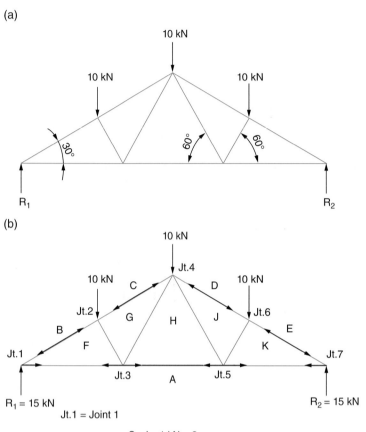

(b)

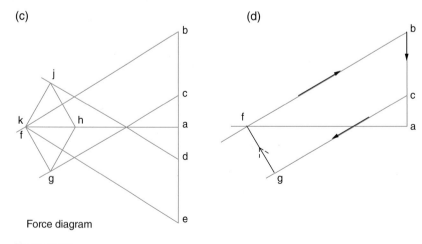

Scale: 1 kN = 2 mm

Force diagram

Figure 10.36

Solution:

1) As the truss is symmetrically loaded, $R_1 = R_2$, and:

$$R_1 + R_2 = \text{total force acting on the truss}$$
$$R_1 + R_2 = 30\,\text{kN}, \text{therefore } R_1 = R_2 = 15\,\text{kN}$$

2) Label the space diagram with capital letters, as shown in Figure 10.36(b). Members BF, CG, DJ and EK are in compression, and members FA, HA and KA are in tension. The directions of forces in these members are shown in Figure 10.36(b). The nature of forces in the remaining members is to be determined.

3) The joints are numbered as shown. Select an appropriate scale to draw the force diagram, say 1 kN = 2 mm.

4) Start from joint 1. Using a scale of 1 kN = 2 mm, draw line ab vertically up to represent force AB (R_1). Force AB is 15 kN, therefore ab should be 30 mm long (Figure 10.36(c)).

5) Go clockwise at joint 1 to the next member, i.e. BF. From point b, draw a line parallel to member BF to represent the force in this member. The direction of line bf should be the same as the direction in which the force in member BF acts at joint 1, i.e. downwards. Since the force in member BF is not known, point f cannot be established as yet.

6) The next member is FA. As the member is horizontal, draw a horizontal line that passes through point a and thus establishes point f. This completes the process at joint 1.

7) The next joint to be considered is joint 2. At this joint, there are three members of the truss and an external force of 10 kN. On the force diagram, mark bc = 20 mm to represent the downward-acting vertical force of 10 kN (1 kN = 2 mm, therefore 10 kN = 20 mm).

8) Go clockwise at joint 2 to the next member, i.e. member CG. To represent the force in member CG at joint 2, draw a line from point c that slopes downwards and is parallel to member CG. Since the force in member CG is not known, point g cannot be established at this stage.

9) Moving clockwise at joint 2, the next member is GF. The magnitude, as well as the direction of the force, in this member is not known. To represent the force in member GF, draw a line on the force diagram that is parallel to member GF and passes through point f. This will establish point g, as shown.

10) The next member is FB, which has already been represented on the force diagram. In order to determine whether there is tension or compression in member GF, let us look at the force diagram from joint 2 (Figure 10.36(d)). The known directions are marked with bold arrows. The arrows must follow each other around the force diagram. Therefore, the direction of the force in member GF at joint 2 is upwards, as shown by the dotted arrow. This shows that there is compression in member GF.

11) The next joint to be considered is joint 3. At this joint, there are four members, two of which are horizontal. Members AF and FG have already been represented on the force diagram. From point g, draw a line parallel to member GH. The direction of the force in member GH is to be determined.

12) The next member in a clockwise direction at joint 3 is HA. As the member is horizontal, the line that represents the force in this member must be horizontal. That line is ha.

13) The nature of the force in member GH can be determined as explained in Step 10. There is tension in member GH.

14) The procedure explained above should be followed at joints 4, 5, 6 and 7 to complete the force diagram.

Table 10.1

Member/s	Compression/tension	Strut/tie	Force (kN)
BF, KE	Compression	Strut	30.0
FA, AK	Tension	Tie	26.0
CG, DJ	Compression	Strut	25.0
GF, KJ	Compression	Strut	8.7
GH, JH	Tension	Tie	8.7
HA	Tension	Tie	17.3

15) To determine the force in each member of the truss, we need to measure the sides of the force diagram and convert them back into kilonewtons.
16) The results are shown in Table 10.1.

Exercise 10.1

1 Calculate the moment about point x for the cantilevers shown in Figure 10.37.

(a)   (b)

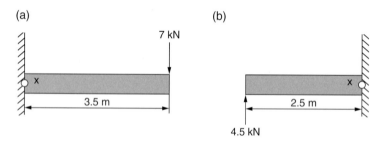

Figure 10.37

2 Calculate the clockwise and anti-clockwise moments about points B and A for the beams shown in Figure 10.38.

(a)   (b)

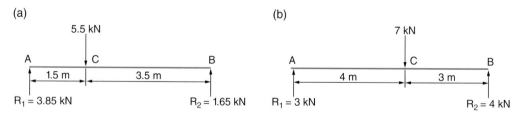

Figure 10.38

3   Figure 10.39 shows the line diagrams of four beams subjected to a variety of loading. Calculate reactions R_1 and R_2.

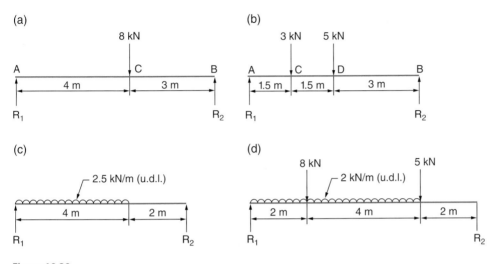

(a)
(b)
(c)
(d)

Figure 10.39

4   For the beams shown in Figure 10.39, produce shear force and bending moment diagrams.

5   A weight of 30 N is supported by two strings, as shown in Figure 10.40. Determine, graphically, the nature and magnitude of the forces in the two strings.

Figure 10.40

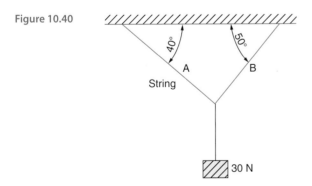

6   Two rods, A and B, support objects weighing 600 N, 300 N and 400 N, as shown in Figure 10.41. For each case, determine the nature and magnitude of the forces in the two rods.

(a)   (b)   (c)

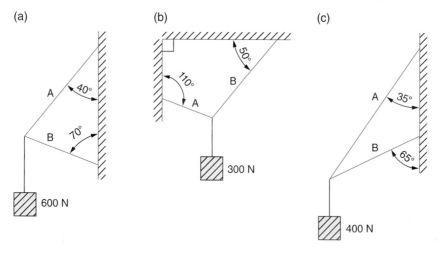

Figure 10.41

7   Find the magnitude and the nature of the forces in members of the trusses shown in Figure 10.42.

(a)   Figure 10.42

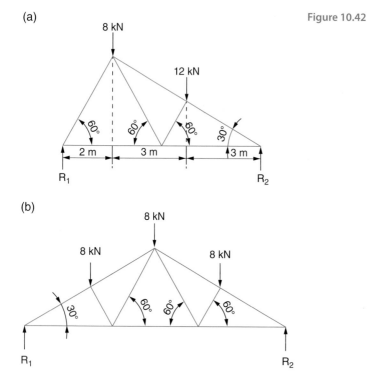

(b)

References/Further Reading

1 Durka, F., Al Nageim, H., Morgan, W. and Williams, D. (2002). *Structural Mechanics*. Harlow: Prentice Hall.
2 Hulse, R. and Cain, J. (2003). *Structural Mechanics*. Basingstoke: Palgrave Macmillan.

11

Forces and Structures 3

LEARNING OUTCOMES

1) Explain the structural behaviour of beams, columns, slabs and foundations.
2) Explain the behaviour of R.C./steel/timber beams.
3) Explain the modes of failure in structural elements.

11.1   Introduction

In Chapters 9 and 10, the basics of structural mechanics, such as tension, compression, types of loading, bending moments, shear forces etc., were discussed. In this chapter, building upon what was discussed in the previous two chapters, we shall focus on more in-depth structural behaviour of beams, columns and foundations. The discussion here is brief, as this subject is vast, and readers should consult the references given at the end of the chapter if they need more information on this topic.

11.2   Beams

11.2.1   Tension and Compression in Beams

Beams are used in a number of situtions in buildings, some of which are:

- To support the weight of brickwork/blockwork above doors and windows;
- To form part of structural frames for high-rise buildings;
- To strengthen the foundations.

The engineer must design a beam in such a way that it has adequate strength to resist bending moments and shear forces, the deflection does not exceed the permissible limit and it can resist lateral buckling. Consider a vertical component acted upon by a tensile force at its centre, and a simply supported beam, as shown in Figure 11.1.

The vertical component is subjected to direct tension whereas the beam is subjected to compression (C) and tension (T) due to bending, and both of them will offer resistance to the external forces, as illustrated. The compressive and tensile stresses in the beam vary as shown, the maximum compressive stress occuring at the top and the maximum tensile stress occuring at the bottom of the beam. There

Construction Science and Materials, Second Edition. Surinder Singh Virdi.
© 2017 John Wiley & Sons Ltd. Published 2017 by John Wiley & Sons Ltd.
Companion website: www.wiley.com/go/virdiconstructionscience2e

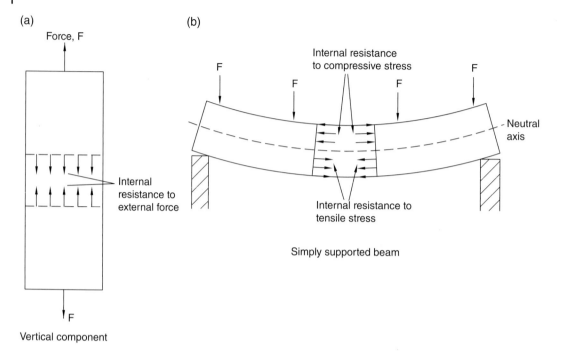

(a)

Force, F

Internal resistance to external force

F

Vertical component

(b)

F   Internal resistance to compressive stress   F

F   F

Neutral axis

Internal resistance to tensile stress

Simply supported beam

Figure 11.1

is no stress at the neutral axis and this is the reason why, in steel beams, most of the material is concentrated in the flanges for optimum resistance to bending (see Figure 11.11).

11.2.2 Shear

Forces acting on beams cause vertical as well as horizontal shear stress. In **vertical shear**, the forces tend to shear the beams from their ends (Figure 11.2). **Horizontal shear** can be demonstrated by placing a number of thin strips of a smooth material on top of each other, as shown in Figure 11.3(a). When a force is applied, the strips bend as they slide over one another (Figure 11.3(b)). In a solid beam this cannot happen as there are no layers; however, the tendency to sliding is present and because of this tendency shear stresses are developed.

Vertical and horizontal shear can also be explained by imagining a small cube of the beam material (Figure 11.4(a)). When a load is applied, vertical shear forces v act on the sides of the cube, as shown. These are set up due to the reaction and force F. These two forces would tend to rotate the cube in a clockwise direction. Since the cube in the beam cannot turn, it must be acted upon by horizontal shear forces h, which set up anti-clockwise rotation. These forces (forces h) tend to cause horizontal shear in a timber joist.

Forces h and v, which act on the cube, can be combined by the parallelogram of forces law to give a resultant force (R) that tends to break the cube in tension (Figure 11.4(b)). As corners B and D move away from each other, causing tension, corners C and A tend to come closer, causing compression. It can, therefore, be concluded that horizontal shear stresses in beams accompany vertical shear stresses which can produce compressive and tensile stresses on diagonal planes. Figure 11.5 shows the failure of a beam due to shear if the material is weak in tension. If the beam material is weak in compression, then the failure planes will be in the other direction.

(a)

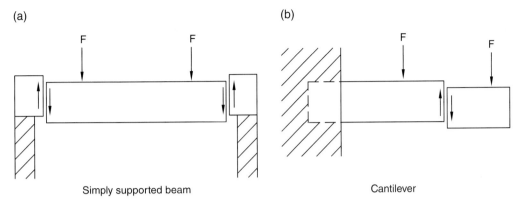

Simply supported beam

(b)

Cantilever

Figure 11.2 Vertical shear in beams.

(a)

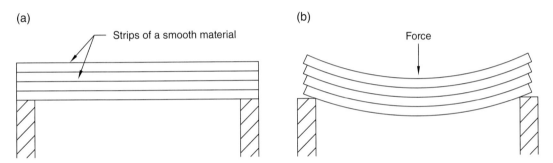

Strips of a smooth material

(b)

Force

Figure 11.3 Horizontal shear in beams.

(a)

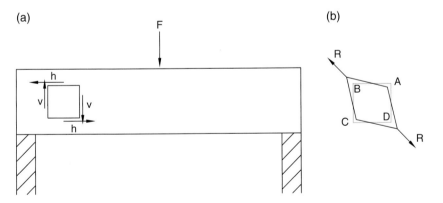

(b)

Figure 11.4 Horizontal and vertical shear in beams.

11.2.3 Deflection

A beam may be satisfactory in terms of resisting bending moments and shear force and yet be unsuitable if its deflection is excessive. Excessive deflection causes cracking of floor and ceiling finishes and, generally, people would not feel safe in a building where this had happened. The deflection of a beam depends on the magnitude of loading, the type of beam supports, beam size and span, and the

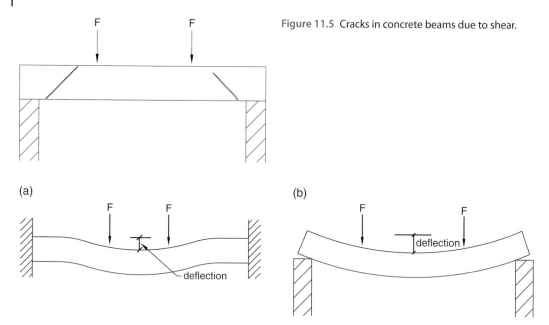

Figure 11.5 Cracks in concrete beams due to shear.

(a)

(b)

Figure 11.6 Deflection.

material from which it is made. If we compare two beams, which are similar except for the end conditions, the beam with fixed ends (Figure 11.6(a)) will deflect less than the simply supported beam (Figure 11.6(b)).

Deflection due to a central point load (W) is given by:

$$\text{Deflection} = \frac{WL^3}{48EI}$$

The effect of span on deflection is very important. Consider 3 m long and 6 m long steel beams with the same cross-section and subjected to the same amount of loading. The value of $\frac{W}{48EI}$ from the above formula will be the same for both beams; therefore, we can conclude that the deflection is proportional to L^3. The 6 m long beam will deflect 8 times more than the 3 m long beam.

The material from which a beam is made is an important factor that has to be taken into account and this is why the modulus of elasticity (E) is used in the deflection formulae. The stiffness of a beam, and hence its resistance to bending, increases as the value of E increases.

Variations in shape and size, especially the depth, cause variations in stresses in the material and thus in the strains. The greater the depth of a beam for a given cross-sectional area, the stiffer it will be. The stiffness of a beam also depends on its second moment of area, I (also called the **moment of inertia**), which is the geometrical property of the shape and size of a particular section irrespective of the material used. The greater the value of I, the stiffer will be the beam.

11.2.4 Lateral Buckling

It has been stated earlier that the depth of a beam is very important in resisting a given bending moment and also to keep deflections small. However, if the depth is too great as compared to the

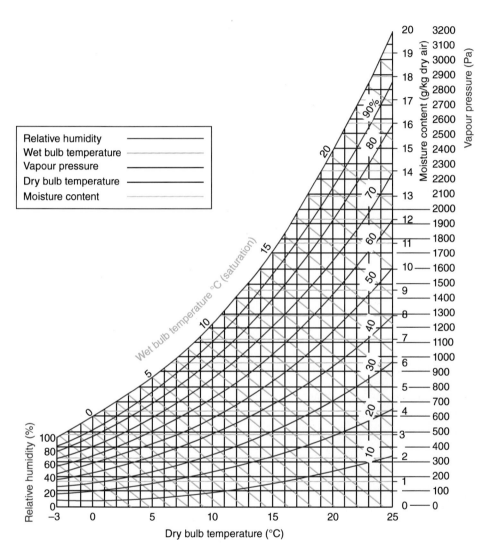

Figure 8.12 The psychrometric chart.

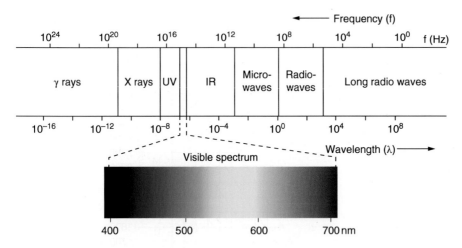

Figure 14.1 Electromagnetic spectrum.

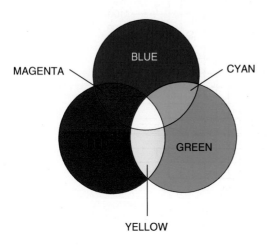

Figure 14.3

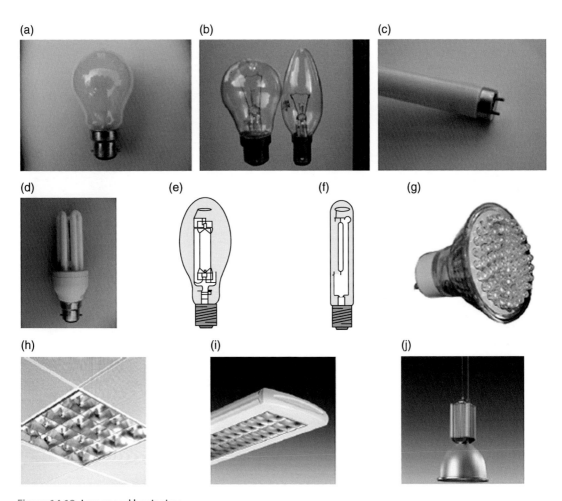

Figure 14.12 Lamps and luminaires.

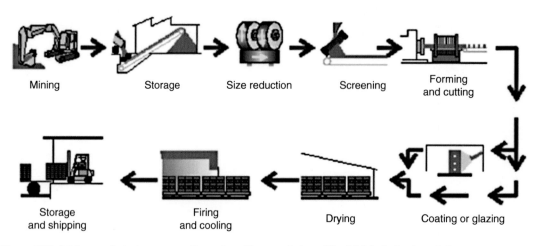

Mining

Storage

Size reduction

Screening

Forming
and cutting

Storage
and shipping

Firing
and cooling

Drying

Coating or glazing

Figure 16.1 Brick manufacturing process. Reproduced by permission of the Brick Industry Association.

Figure 16.2 Brick spalling.

Figure 16.3 Brick efflorescence.

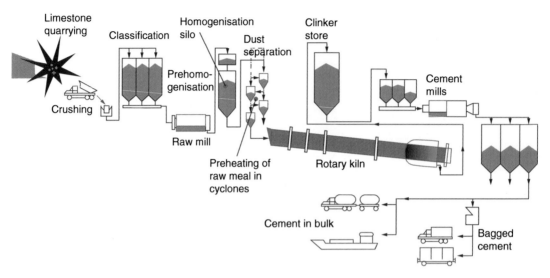

Figure 16.4 Cement manufacturing process. Reproduced by permission of Kääntee, U.

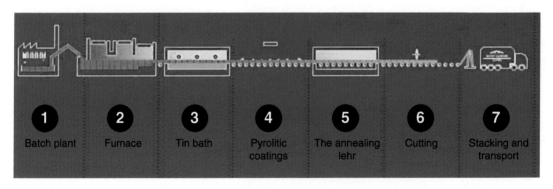

Figure 16.8 Glass manufacturing process. Reproduced by permission of Saint-Gobain, UK.

Figure 11.7 Lateral buckling in beams.

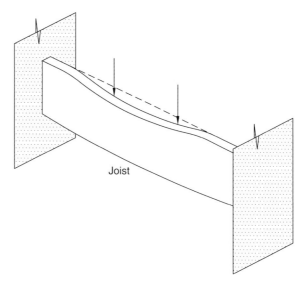

Joist

width, the beam may buckle sideways due to the 'column effect' (see Figure 11.7). The compressive forces in the upper fibres of a beam have the same effect as those in a column and the tendency to buckle increases as the beam becomes deeper and thinner (see Section 11.7.1 for details).

When a beam is embedded in a reinforced concrete floor, there is sufficient lateral restraint, but when the beam is not restrained laterally, measures need to be taken to counteract the sideways movement.

11.3   Reinforced Concrete (R.C.) Beams

Fresh concrete is semi-fluid, and when poured into a mould it takes the shape of that mould. The concrete becomes so hard and strong after a few days that it behaves like stone, and like natural stone, concrete is strong in compression but weak in tension. As beams are designed to carry load, concrete is reinforced with steel bars to resist tensile stress. Tensile stress normally occurs below the neutral axis, but depending on the end supports/restraints, it may also occur above the neutral axis.

Steel bars are stretched due to tension and although the extension is minute, it is sufficient to cause cracking of concrete in the tension zone (the cracks are extremely small). As the concrete cracks in the tension zone, it is unable to resist the tension; however, due to a strong bond between the concrete and the steel bars, the tension is resisted by the steel, as shown in Figure 11.8.

Figure 11.9 shows typical reinforcement in three beams to resist tensile stresses. More steel is provided to resist thermal stresses and for the fabrication of the reinforcement cage.

11.3.1   Shear Reinforcement

If the magnitude of shear stress in a concrete beam is very high, it could cause failure, as shown in Figure 11.5. The reinforcement provided in a beam to resist the tensile stress due to bending (Figures 11.8 and 11.9) does not help the concrete to resist diagonal tensile stress. Reinforcement (i.e. shear steel) must be provided to resist diagonal tension; this can take the form of bent-up bars or vertical stirrups or a combination of both (Figure 11.10).

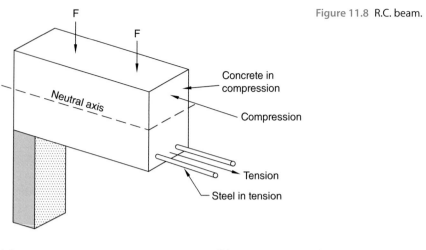

Figure 11.8 R.C. beam.

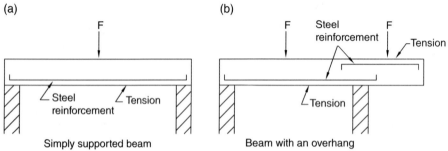

(a)

Simply supported beam

(b)

Beam with an overhang

(c)

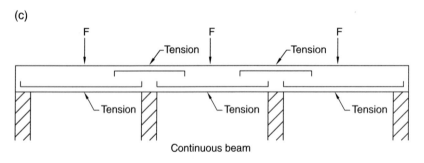

Continuous beam

Figure 11.9 Steel reinforcement to resist tensile stress.

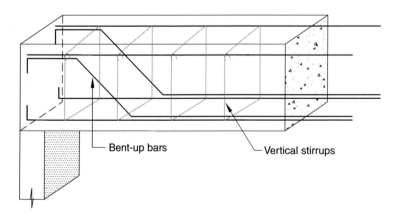

Figure 11.10 Shear reinforcement.

11.4   Steel Beams

Steel is strong in compression and in tension, economical and easily available in large quantities. One of the main applications of steel in construction is the production of structural sections such as universal beam/column sections, angle sections, channels and hollow sections (see Figure 11.11).

The shape of a universal beam section involves the concentration of the material at the flanges because:

- The material near the top and bottom of the beam is subjected to higher stresses as compared to the material near the neutral axis (also explained in Section 11.2.1).
- The **lever arm** of an I-section would be much greater than that for the beam of a rectangular section; hence, its moment of resistance (MoR) would be greater than the MoR of a rectangular section. The explanation is given below.

In a rectangular section, the tensile as well as the compressive stress is assumed to act at d/3 from the neutral axis. Thus, the lever arm is 2d/3 (Figure 11.12), which is smaller than the lever arm when the stresses are assumed to act at the outer edges. For this reason, the maximum amount of material in a steel beam is provided as far as possible from the neutral axis; this creates the typical I-shape of universal beam sections.

11.4.1   Bending

As explained previously, tensile and compressive stresses are set up in a steel beam due to bending. These stresses are small near the neutral axis but much larger in the extreme fibres. As the load is

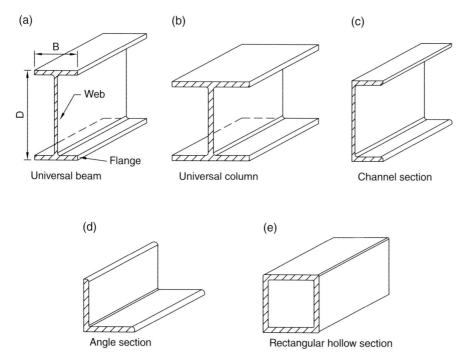

Figure 11.11 Structural steel sections.

(a)   (b)

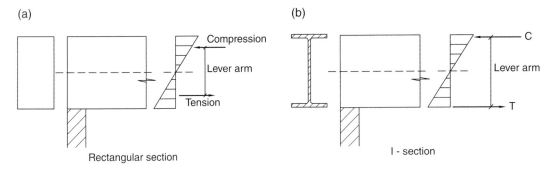

Figure 11.12 Lever arm in beams.

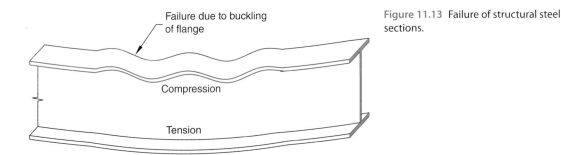

Figure 11.13 Failure of structural steel sections.

increased further (and hence the bending moment), failure will occur eventually when the steel yields in compression and/or tension. If the flanges and the web are too thin, the beam may fail by crushing of the metal that is in compression and/or tearing of the metal that is in tension (Figure 11.13). This failure is due to the formation of a plastic hinge at the point of the maximum bending moment.

11.4.2 Plastic Hinge

Consider a simply supported beam with a small point load W; the moment at the mid span is Wl/4. The stress distribution is shown in Figure 11.14(b), and when the load is removed, the beam reverts to its original shape and size (Figure 11.14(e)). As the load is increased, the bending moment increases and the stresses in extreme fibres reach the yield stress (Figure 11.14(c)). As the load continues to increase, the remaining fibres of the beam reach the yield stress and eventually the whole cross-section reaches the yield stress (Figure 11.14(d)). The moment corresponding to this state is known as the **plastic moment** (M_p). Any attempt to increase the moment beyond M_p will cause the beam to act as if it is hinged, as shown in Figure 11.14(f). This is known as a **plastic hinge**, which allows significant rotation to occur.

11.4.3 Shear

A steel beam may fail in shear if the shear forces are excessive. The web of a beam resists shear force, but if it is too thin, failure may take place due to buckling of the web, as shown in Figure 11.15. This problem may arise in deep beams (for example, plate girders) and it is important to strengthen the web, for example by fixing angle sections.

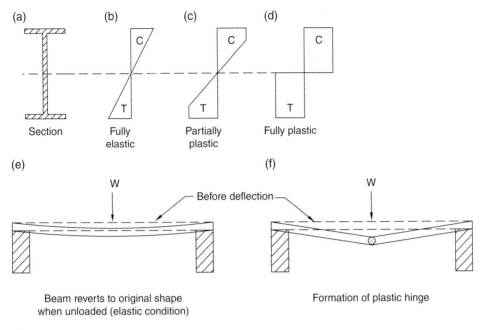

Figure 11.14

Figure 11.15 Failure due to shear.

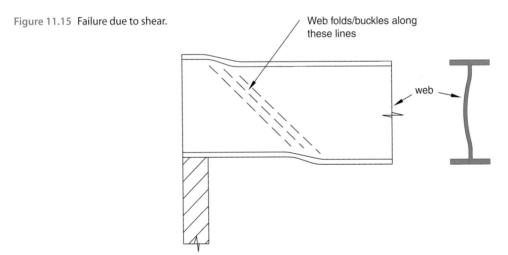

11.5 Timber Joists

Timber is one of the most important materials used in construction as it is strong, lightweight, aesthetically pleasing, durable and sustainable. There are two types of timber – softwood and hardwood – softwoods being used mainly for structural elements/components such as trussed rafters, joists, floor boards etc.

The structural behaviour of beams, discussed in Section 11.2, applies to timber joists as well. However, unlike other materials, timber has a grain; therefore, its strength varies in different planes, as illustrated in Figure 11.16.

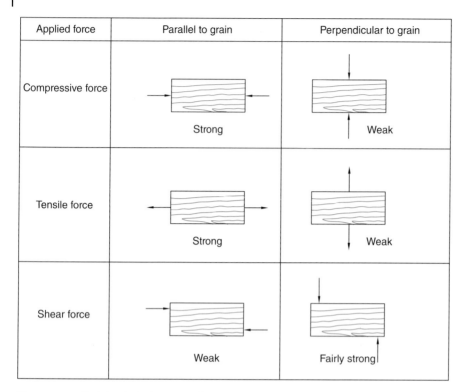

Figure 11.16 Variation in timber strength.

11.5.1 Failures in Timber Joists

Compression/Tension Failure

If the load on a joist is excessive it will break or fail; these failures may be classified according to the way in which they develop – as tension, compression and horizontal shear – and according to the appearance of the broken surface.

The tensile strength of timber is, on average, about two to three times as great as the compressive strength. A beam should, therefore, be expected to fail initially by the formation of a fold on the compression side due to the crushing of fibres, followed by failure on the tension side. This is usually the case in either moist wood or high-strength timber. In seasoned or dry timber, the stiffness of the fibres increases, which may not have much effect on the tensile strength but enhances the resistance to crushing. For this reason, it is more common to find that the first visible failure is usually on the lower or tension side. There is considerable variation in tension failures, caused by the toughness of the wood, the slope/position of the grain, knots and other defects. Some of the common forms of failure are:

1) The tensile force, acting parallel to the grain, pulls the timber in two, as shown in Figure 11.17(a).
2) If the joist has diagonal grain, the tensile force acts obliquely to the grain. Failure occurs as the tensile strength of timber across the grain is only a small fraction of that with the grain, as shown in Figure 11.17(b).
3) Figure 11.17(c) shows the failure mode which is most likely to occur. The cause is a combination of tension and shear; however, it is started off by a tension fracture.

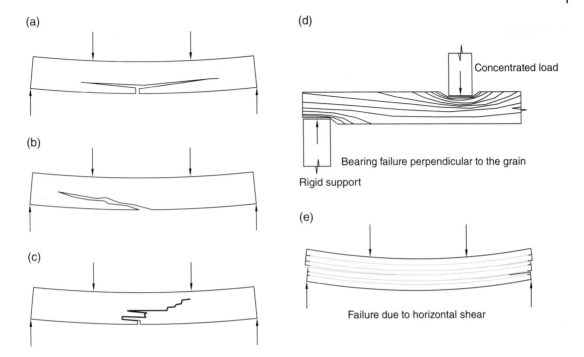

Figure 11.17 Failure in timber joists.

Bearing Failure

When timber is compressed perpendicular to the grain, the wood fibres can withstand the load initially, but as the load becomes excessive, the fibres are fully squashed and failure occurs, usually by shearing across the grain (see Figure 11.17(d)).

Horizontal Shear Failure

Horizontal shear was explained in Section 11.2.2 by considering a number of strips and their sliding when external forces were acting on beams. As a tree trunk gets bigger every year due to the formation of sapwood, a timber joist may be considered to be made of naturally bonded layers of wood, and hence analogous to Figure 11.3.

The upper and lower portions of the joist slide along each other for a portion of their length either at one or at both ends, as shown in Figure 11.17(e). Failure is often due to a defect in timber, which reduces the actual area resisting the shearing force considerably.

11.5.2   Lateral Buckling

The depth of a timber joist is usually much greater than its width. For this reason and due to the column effect in beams, discussed in Section 11.2.4, the joists are susceptible to lateral buckling. The floor can be strengthened and the movement of the joists checked by providing lateral support, i.e. solid strutting or herring-bone strutting; the former is shown in Figure 11.18.

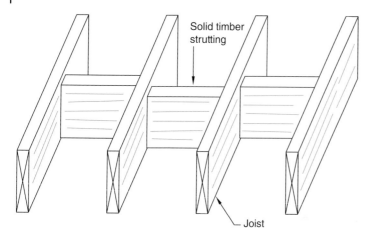

Figure 11.18 Solid timber strutting.

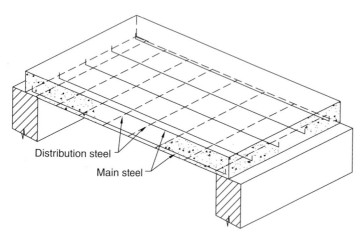

Figure 11.19 One-way concrete slab.

11.6   Slabs

In commercial or multi-storey residential buildings, the floors may be constructed from either cast *in situ* (at site) concrete or precast concrete units. The concrete slab may be designed as a one-way slab, where the slab is supported on just two supports, as shown in Figure 11.19.

The bulk of the steel is provided along the span, but a smaller quantity of steel, known as the **distribution steel**, is provided at right angles to the span. As fresh concrete starts to gain strength, secondary stresses develop due to temperature changes and shrinkage; the distribution steel resists the secondary stresses.

If the length of a room is not much different from its width, the concrete slab may be designed as a two-way slab, assuming that it is supported on all four edges. In this slab, the main steel will span in both directions; the area of steel, and hence the number of reinforcing bars, is calculated by performing appropriate structural calculations. The thickness of a two-way slab is less than that of a similar one-way slab. Precast concrete beams, which are available in different sizes and formats, may be used for floor construction if the size of the job justifies it.

(a)

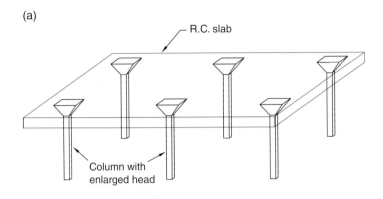

(b)

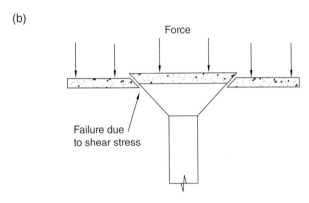

Figure 11.20 Flat slab floor.

If the head-room in a building is reduced due to the use of deep beams, then a flat slab floor may be provided. The flat slab requires no beams and is supported on columns, as shown in Figure 11.20(a). The columns may be provided with enlarged heads to reduce shear stresses. The concrete has to fail around the perimeter of the column head if a failure due to shear is to take place (Figure 11.20(b)).

11.7 Columns

A column is a vertical structural member used in buildings, bridges and other structures. In buildings, columns may support a roof or a beam or they can be purely decorative. Columns may be constructed using a range of materials such as bricks, stone, timber, concrete and metals. Cast iron columns were used in many buildings during the 19th century but due to the low tensile strength of this metal, reinforced concrete and mild steel have become the main materials for columns in modern construction.

11.7.1 Slenderness Ratio

The behaviour of a column depends on its height relative to its thickness. The column is called a short column if its height, relative to its thickness, is small. The load-carrying capacity of a short column depends on the material from which it has been made. If a column is too high as compared to its thickness, the column is called a tall, or slender, column. The tall column will become unstable by buckling at a load that may not cause any damage to a short column of the same size and material.

The tendency of a column to buckle depends on its height and thickness; the ratio between the two is known as the **slenderness ratio**:

$$\text{Slenderness ratio} = \frac{\text{Effective height of the column}}{\text{Least radius of gyration of the column}}$$

If the slenderness ratio is high, the tendency of the column to buckle will be high and vice versa. The buckling will occur in the direction of least thickness, i.e. in the direction of the least resistance.

Consider two columns, A and B, with the same height and the same cross-sectional area (Figure 11.21).

$$\text{Slenderness ratio of column A} \quad = \quad \frac{4000}{200} = 20$$

$$\text{Slenderness ratio of column B} \quad = \quad \frac{4000}{100} = 40$$

Due to its higher slenderness ratio, column B will buckle under a smaller load as compared to column A.

11.7.2 Effective Height of Columns

It was explained in Section 11.2.3 that beams fixed at both ends behave differently from simply supported beams. Similarly, the end conditions of a column affect its load-carying capacity; these end conditions are shown in Figure 11.22. A column is free to rotate if its end is hinged, whereas the tendency to rotate is restrained if the end is fixed.

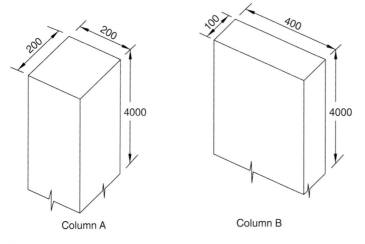

Column A

Column B

Figure 11.21

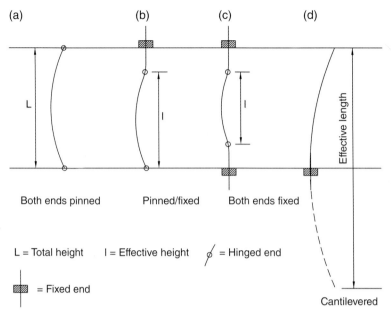

(a) (b) (c) (d)

L

Effective length

Both ends pinned   Pinned/fixed   Both ends fixed

Cantilevered

L = Total height   I = Effective height   ⌀ = Hinged end

▨ = Fixed end

Figure 11.22 Effective length of columns.

Table 11.1 Effective length for compression members.

End condition	Effective length
Both ends pinned	L[*]
Both ends fixed	0.7 L
One end fixed and one pinned	0.85 L

[*] L is the actual length of the column

It can be seen from Figure 11.22 that column C buckles less when compared to the other columns. The reason for this behaviour is that the effective length of column C will be the smallest, and hence its slenderness ratio and tendency to buckle will be the lowest. Table 11.1 shows the effective length of columns for different end conditions.

The shape of the section also affects the value of the safe column load. The cross-sectional dimensions are taken into account by considering the **radius of gyration** (r), which is given by:

$$r = \sqrt{\frac{I}{A}}$$

where I is the second moment of area and A is the cross-sectional area.

For the section shown in Figure 11.23(a):

$$r_{xx} = \sqrt{\frac{I_{xx}}{A}}$$

$$r_{yy} = \sqrt{\frac{I_{yy}}{A}}$$

The radius of gyration is the distance to a point or axis at which the whole material of the column is assumed to be concentrated from the axis of buckling.

11.7.3 Eccentric Loading on Columns

A column may carry axial (concentric) or eccentric load. If the resultant load acts at the centre of gravity of the column, it is known as an **axial load**. The column is thus subjected to only compressive stress, which is known as direct compressive stress. If the resultant load does not act at the centre of gravity of the column, then it is known as an **eccentric load**, as shown in Figure 11.23.

An axial load causes compressive stress in a column whereas an eccentric load causes compressive stress and sets up bending stresses, i.e. compressive stress on one side and tensile stress on the other. If the eccentricity is small (Figure 11.23(b)), then the compressive stress will increase on side A and reduce on side B, as the tensile stress and compressive stress are opposite to each other. If the eccentricity of the load is large (Figure 11.23(c)), then the compressve stress on side A will increase, but on side B the tensile stress may be large enough to cancel out the compressive stress. This could result in (net) tensile stress on side B, which could be dangerous in materials with low tensile strength such as bricks and unreinforced concrete.

Figure 11.23 shows three columns, each subjected to load W. If A represents the cross-sectional area of the column and e is the eccentricity of the load, then:

$$\text{Column a: Compressive stress} = \frac{W}{A}$$

$$\text{Column b or column c: Stress on face A} = \frac{W}{A} + \frac{We}{Z} \left(\text{compressive stress}\right)$$

$$\text{Stress on face B} = \frac{W}{A} - \frac{We}{Z} \left(\text{compressive or tensile stress}\right)$$

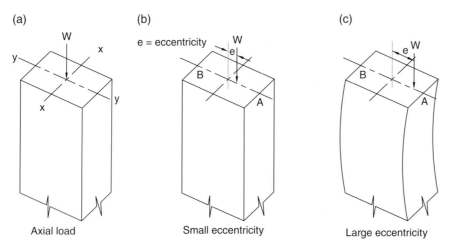

Figure 11.23 Eccentricity of column load.

Z in the above formulae is called the **section modulus** and for a solid rectangular section, b wide and d deep, it is given by:

$$Z = \frac{bd^2}{6}$$

The large eccentricity of loading, shown in Figure 11.23(c), may cause tensile stress in face B of the column. When designing unreinforced concrete/brickwork columns, the engineer has to make sure that tension does not develop in these materials, or the calculation $\frac{W}{A} \pm \frac{We}{Z}$ gives positive stresses. Refer to Figure 11.23(c), and remembering that compressive stresses are positive, the stress along face B can be written as:

$$\text{Stress} = \frac{W}{A} - \frac{We}{Z} > 0$$

$$\text{or} \quad \frac{W}{A} \geq \frac{We}{Z}$$

$$\text{Therefore, } e \leq \frac{Z}{A}$$

For a solid rectangular section, b wide and d deep, the above inequality becomes:

$$e \leq \frac{bd^2/6}{bd} \quad \text{Therefore } e \leq \frac{d}{6}$$

Therefore, provided that load W acts at an eccentricity of d/6 or less from the centroidal axis, no tensile stress will occur along face B.

11.7.4 Steel Columns

Figure 11.24 shows some of the steel sections that can be used in the construction of columns. For steel columns, an H-section, in which the width of the flange is equal to the depth, is more efficient than an I-section with an equal amount of material. The reason is that in columns, resistance to bending is required about the x–x axis as well as the y–y axis.

(a)   (b)   (c)

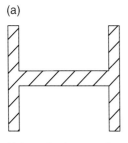 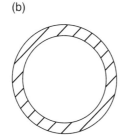 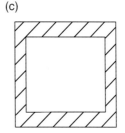

Universal column section   Hollow circular section   Hollow rectangular section

Figure 11.24 Structural steel sections for columns.

(a)

(b)

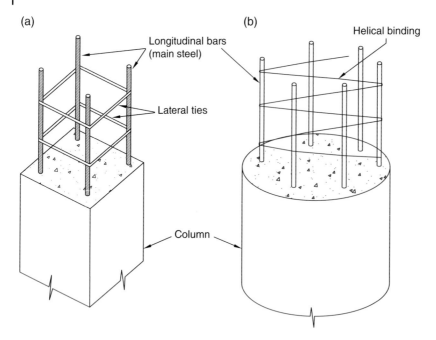

Longitudinal bars (main steel)

Helical binding

Lateral ties

Column

Figure 11.25 Typical reinforcement in R.C. columns.

Hollow sections allow the material to be provided at greater distances from, and symmetrically to, the axes of the column. Hollow steel sections are therefore more efficient than the other sections provided that the walls are not so thin as to allow local crumpling.

11.7.5 Reinforced Concrete Columns

For any particular building, the cross-sections of reinforced concrete columns are larger than the corresponding steel columns. Most R.C. columns in buildings have low slenderness ratios and thus are classified as short columns for buckling. This also means that failure due to buckling of the column is not common.

The area of longitudinal steel is usually between 0.8% and 8%, and since steel is stronger than concrete, the load carried by a column will be directly proportional to the percentage of steel. Figure 11.25 shows typical reinforcement in two columns.

11.8 Foundations

Whether we are talking about a building or a bridge or a dam, its weight is finally transmitted to the ground, i.e. rocks/subsoils. Rocks may have fairly high strength, but subsoils have limited strength; therefore, an engineer has to make sure that the stress in rocks/subsoils, due to the weight of the structure, is within permissible limits. The structural element with which this can be achieved is called the **foundation**. The main functions of a foundation are to ensure that:

- The soil does not fail due to the weight of the building/structure;
- Excessive settlement does not occur;
- Differential settlement, which can cause cracks in a building, does not occur.

11.8.1 Strip Foundation

The force acting on the subsoil due to the weight of a house is usually small; therefore, a narrow strip constructed of unreinforced concrete is usually provided. The typical width of the narrow strip is about 600 mm, but its thickness is chosen in such a way that the foundation does not extend beyond the shear planes. The **shear planes** are imaginary lines within which the force of the building is supposed to be distributed (Figure 11.26). The stress caused by the building load must be balanced by the soil reaction, otherwise there will be excessive settlement of the building.

The load-bearing walls carry a heavy load and tend to cause foundation failure by shear, as shown in Figure 11.27. The thickness of the concrete strip must be adequate to resist this shear stress.

If a wall carries a very heavy load, then a foundation which is wider than the narrow strip may be required to ensure that the stress in the subsoil does not exceed the safe bearing capacity of the soil. The foundation is subjected to bending due to the downward load and the resulting upward reaction from the soil (Figure 11.28). The foundation may be designed as a plain concrete strip if the shear

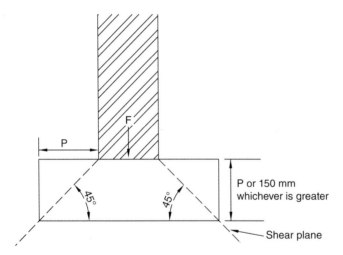

Figure 11.26 Strip foundation.

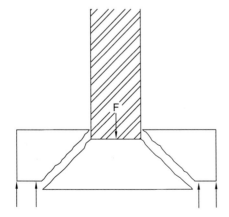

Figure 11.27 Failure in shear.

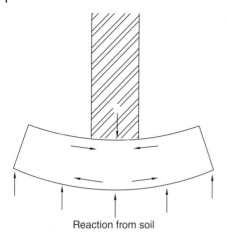

Figure 11.28 Bending in concrete strip foundation.

Reaction from soil

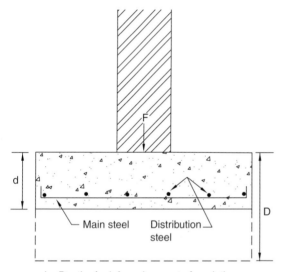

Figure 11.29 Wide strip foundation.

d = Depth of reinforced concrete foundation
D = Depth if the foundation were designed as
    unreinforced section

planes pass through the corners of the strip. The resulting thickness of the plain concrete wide strip may be excessive, and therefore uneconomical. An economical alternative is a thinner reinforced concrete strip; these two options are illustrated in Figure 11.29.

11.8.2 Pad Foundation

The simplest and most economical solution for providing the foundation of a column is to construct a reinforced/unreinforced concrete pad. The column should be constructed centrally on the foundation to avoid eccentricity of loading and hence uneven pressure on the soil. If a column supports a small load, then an unreinforced pad may be satisfactory, but for heavier loads, the foundation

(a) (b)

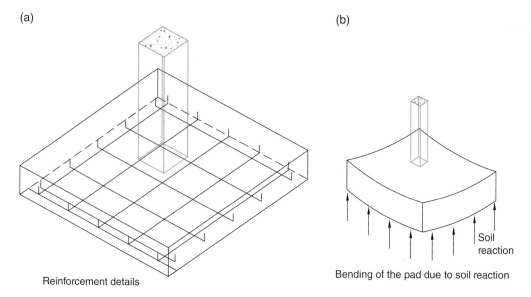

Reinforcement details

Bending of the pad due to soil reaction

Soil
reaction

Figure 11.30 Pad foundation.

must be designed as a reinforced pad (Figure 11.30(a)). The pad bends due to the column load and the upward-acting soil reaction, as shown in Figure 11.30(b). Therefore, apart from being an economical solution, a reinforced pad has steel reinforcement to resist any tensile stresses that might develop.

The column, due to the load it carries, tends to cause failure by shearing through the foundation. The stress caused by this phenomenon is called **punching shear stress**. The thickness of the pad must be sufficient to resist the stress caused by punching shear.

11.8.3 Other Foundations

When a number of columns are closely spaced, the isolated pads may overlap one another and, in this situation, a continuous concrete strip may be provided. The centre of gravity of the column loads must coincide with the centre of gravity of the foundation to avoid eccentricity and uneven stress distribution in the subsoil.

Sometimes a column may be so close to the site boundary that there is not enough space for the isolated pad. Its foundation may be combined with the foundation of an inner column, resulting in the combined rectangular (Figure 11.31) or combined trapezoidal foundation (Figure 11.32). As with the continuous strip, the centre of gravity of the column loads must coincide with the centre of gravity of the foundation. Figure 11.31(b) shows how the foundation would be deflected due to the combined effect of the applied forces and the soil reaction. The distribution of tensile and compressive stress in the foundation can be visualised easily from this diagram.

For raft and pile foundations, readers should refer to the publications mentioned in the references/further reading list below.

(a)

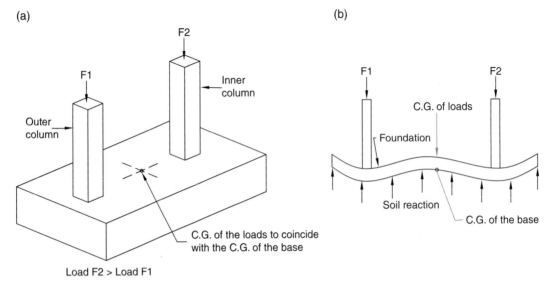

(b)

Figure 11.31 Combined column rectangular foundation.

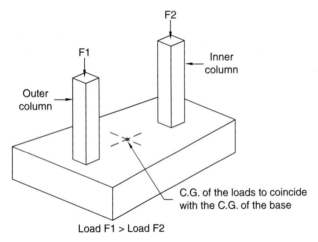

Figure 11.32 Combined column trapezoidal foundation.

References/Further Reading

1 Arya, C. (2009). *Design of Structural Elements: Concrete, Steelwork, Masonry and Timber Design to British Standards and Eurocodes*. Oxford: Taylor and Francis.
2 Seward, D. (2009). *Understanding Structures*. Basingstoke: Palgrave Macmillan.
3 Durka, F., Al Nageim, H., Morgan, W. and Williams, D. (2002). *Structural Mechanics*. Harlow: Prentice Hall.
4 Hulse, R. and Cain, J. (2003). *Structural Mechanics*. Basingstoke: Palgrave Macmillan.

12

Fluid Mechanics

12.1   Introduction

Water, like air, is also necessary for the life of humans, animals and vegetation. However, if it is not controlled properly, water may cause flooding, damage to property and loss of human life. Several types of civil engineering structures, for example, canals, dams, culverts, elevated water tanks, drains etc., are constructed to control/contain water for useful purposes. In all these cases, water exerts pressure on the structure; this needs to be calculated when preparing the structural design. Water also affects the design of retaining walls, building foundations, roofs, cladding and other structures.

In science, both liquids and gases are considered fluids. Although the principles and theories described here are based on water, they can also be applied to other fluids.

In this chapter, some fundamental aspects of pressure exerted by static water (hydrostatic pressure) and pressure exerted by flowing water (hydrodynamic pressure) will be considered.

12.2   Pressure of Fluids at Rest

Fluids exert pressure on the base and the sides of the container in which they are stored. The pressure at any point in a fluid is due to the weight of the fluid above that point. Consider a vessel with a length of L metres and a width of W metres. Let the depth of the liquid in the vessel be h metres (Figure 12.1).

$$\text{Volume of the liquid} = L \times W \times h \quad \left(m^3\right) \tag{12.1}$$

From Chapter 1, Section 1.7, we know that

$$\text{Density}(\rho) = \frac{\text{Mass}(m)}{\text{Volume}(V)}$$

Construction Science and Materials, Second Edition. Surinder Singh Virdi.
© 2017 John Wiley & Sons Ltd. Published 2017 by John Wiley & Sons Ltd.
Companion website: www.wiley.com/go/virdiconstructionscience2e

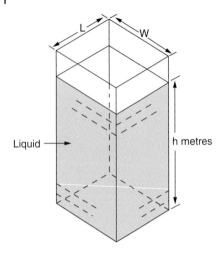

Figure 12.1

Liquid →

h metres

Transposing the above equation,

$$V = \frac{m}{\rho}$$ (12.2)

From Equations (12.1) and (12.2):

$$\frac{m}{\rho} = L \times W \times h \left(m \text{ is the mass of the liquid} \right)$$

After transposition,

$$m = L \times W \times h \times \rho$$
$$= A \times h \times \rho \quad \left(\text{area of the base, } A = L \times W \right)$$ (12.3)

$$\text{Weight of a substance} \left(w \right) = m \times g, \quad \text{or } m = \frac{w}{g}$$ (12.4)

From Equations (12.3) and (12.4):

$$\frac{w}{g} = A \times h \times \rho$$

$$\text{or } w = A \times h \times \rho \times g$$

The force at the base of the vessel is due to the weight of the liquid.

$$\text{Force at the base} = A \times h \times \rho \times g$$
$$\left(\text{Units: newtons} \left(N \right) \right)$$

We also know that:

$$Pressure\,(p) = \frac{Force\,(F)}{Area\,(A)}$$

$$= \frac{A \times h \times \rho \times g}{A} = h \times \rho \times g$$

$$\left(Units : N/m^2 or\ Pascal\ (Pa)\right)$$

The above equation shows us that the pressure exerted by a liquid is proportional to its depth and density. For a particular liquid, the values of density and g remain unchanged; therefore, the pressure depends only on the depth of the liquid. This may be shown by using a tall vessel with a number of side tubes at various depths, as shown in Figure 12.2(a). The vessel is filled with water and the side tubes are unblocked. The speed at which water flows out is the greatest for the lowest jet, showing that pressure increases with depth.

The pressure responsible for the flow from the side tubes acts horizontally; hence, we can say that in liquids, the pressure acts in all directions, not just vertically.

The variation in pressure can be shown as a pressure diagram (Figure 12.2(b)). Owing to the outflow of water, the pressure decreases continuously as the depth of water is never constant. However, when the depth of water is 0.6 m, i.e. the moment when the test starts, the pressures at points A, B and C are:

$$Pressure\ at\ point\ A = h \times \rho \times g \quad \left(g = 9.81\,ms^{-2}\right)$$

$$= 0.2 \times 1000 \times 9.81 \quad \left(density\ of\ water,\ \rho = 1000\,kg/m^3\right)$$

$$= 1962\,Pa$$

$$Pressure\ at\ point\ B = 0.4 \times 1000 \times 9.81 = 3924\,Pa$$

$$Pressure\ at\ point\ C = 0.6 \times 1000 \times 9.81 = 5886\,Pa$$

Figure 12.2 Variation of liquid pressure with depth.

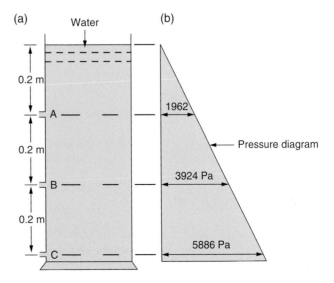

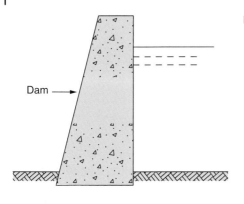

Dam →

Figure 12.3 Concrete dam.

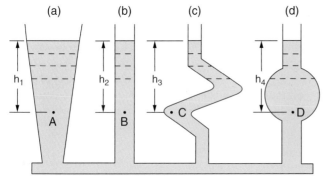

(a)   (b)   (c)   (d)   Figure 12.4

The variation in liquid pressure is used in water-retaining and soil-retaining structures to achieve an economical design. Figure 12.3 shows that dams are made much thicker at the bottom, as the maximal pressure occurs at the base of the reservoir.

The pressure is independent of the shape of the vessel. In Figure 12.4, there are four vessels of different shapes but filled with the same liquid to identical heights. The density of the liquid and the acceleration due to gravity remain the same in each case.

Pressure at point $A\left(p_A\right) = h_1 \times \rho \times g$

Pressure at point $B\left(p_B\right) = h_2 \times \rho \times g$

Pressure at point $C\left(p_C\right) = h_3 \times \rho \times g$

Pressure at point $D\left(p_D\right) = h_4 \times \rho \times g$

If $h_1 = h_2 = h_3 = h_4 = h$, then $p_A = p_B = p_C = p_D = h \times \rho \times g$

The force (or thrust) at points A, B, C or D, due to the weight of the liquid above, will be different as the force depends on the area on which the pressure acts.

12.3   Why do Liquids Flow?

In solids like bricks, steel, timber etc. the molecules are arranged in regular patterns touching each other. The force of attraction between the molecules is so strong that they cannot move. In liquids, the molecules are still close together but not necessarily touching others. The forces of attraction

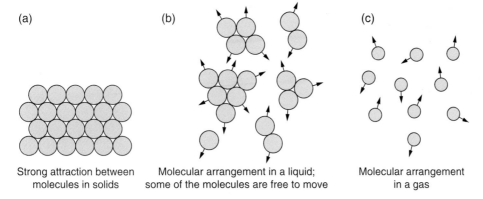

Strong attraction between
molecules in solids

Molecular arrangement in a liquid;
some of the molecules are free to move

Molecular arrangement
in a gas

Figure 12.5 Molecular arrangement (2D) in (a) a solid; (b) a liquid; (c) a gas.

between molecules are therefore weaker than those in solids. The molecules can move about in all directions, hence liquids will flow and take the shape of the container in which they are stored. Figure 12.5 shows the difference between molecular arrangements in solids, liquids and gases.

12.4   Centre of Pressure

Figure 12.2(b) shows how pressure varies from zero at the surface of a liquid to its maximum value (hρg) at the base.

$$\text{Average pressure} = \frac{0 + h\rho g}{2} = \frac{h\rho g}{2}$$

This is the same as calculating pressure at a depth of $\dfrac{h}{2}$ and applying it for the whole depth. Average pressure is used in calculating the total thrust on an object or engineering structure.

The centre of pressure is the point through which the resultant force acts. For a rectangular surface, the centre of pressure is at 2/3 of the immersed depth below the surface, or at 1/3 of the immersed depth measured from the bottom (see Figure 12.6).

Figure 12.6 Centre of pressure.

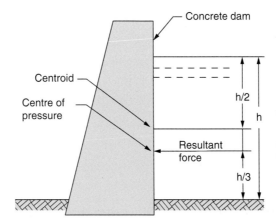

Example 12.1 A 200 m long concrete dam has a vertical face on the reservoir side. Calculate the thrust on the dam if the depth of water is 20 m, and the point of action of this thrust. Density of water $= 1000\,\text{kg/m}^3$ and $g = 9.81\,\text{ms}^{-2}$.

Solution:
Depth of water, $h = 20\,\text{m}$; density, $\rho = 1000\,\text{kg/m}^3$

$$\text{Pressure at the centre of area} = \frac{h\rho g}{2}$$
$$= \frac{20 \times 1000 \times 9.81}{2}$$
$$= 98100\,\text{N/m}^2 \text{ or } 98.1\,\text{kN/m}^2$$

Area on which the water pressure acts $= 200 \times 20 = 4000\,\text{m}^2$

Thrust on the dam $= \text{pressure} \times \text{area}$
$$= 98.1 \times 4000 = 392400\,\text{kN or } 392.4\,\text{MN}$$

Example 12.2 A lock gate[1] is 25 m wide. On one side, the water is 9 m above the lower edge of the gate and on the other side the water is 4.5 m above the lower edge of the gate, as shown in Figure 12.7. Calculate:

a) The resultant force on the lock gate.
b) The point of action of the resultant force.

$$\rho = 1020\,\text{kg/m}^3,\ g = 9.81\,\text{ms}^{-2}.$$

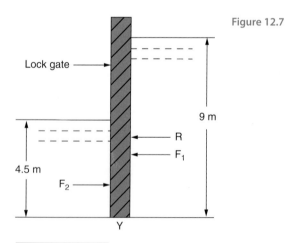

Figure 12.7

---

1 Lock gates are used to lift or lower boats from one level to another. They are found on many canals and rivers.

Solution:

Force $F_1 = p_1 \times A_1$

$$= \frac{9 \times 1020 \times 9.81}{2} \times \left(25 \times 9\right)$$

$$= 1.013 \times 10^7 \, N = 10.131 \, MN$$

This force acts at a distance of $\dfrac{9}{3}$ or 3.0 m from Y.

Force $F_2 = p_2 \times A_2$

$$= \frac{4.5 \times 1020 \times 9.81}{2} \times \left(25 \times 4.5\right)$$

$$= 2.533 \times 10^6 \, N = 2.533 \, MN$$

This force acts at a distance of $\dfrac{4.5}{3}$ or 1.5 m from Y.

Resultant force $\left(R\right) = 10.131 - 2.533 = 7.598 \, MN$

Let us assume that the resultant force acts at a distance of d metres from Y. To find the point of action of the resultant force, take moments about point Y:

$$7.598 \times d = 10.131 \times 3 - 2.533 \times 1.5$$

$$= 26.5935$$

Therefore,

$$d = \frac{26.5935}{7.598} = 3.5 \, m$$

The resultant force on the lock gate is 7.598 MN, which acts at 3.5 m from the base of the gate.

12.5   The Flow of a Fluid

The flow of a fluid may be classified as laminar, transitional or turbulent. In **laminar flow**, the particles move in well-defined paths which are straight and parallel to the pipe walls. The particles do not cross the flow paths of other particles (Figure 12.8(a)). The flow is slower close to the pipe walls due to the friction between the fluid and the pipe material, and faster near the middle of the

(a)   (b)

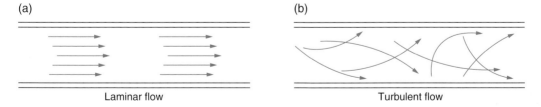

Laminar flow   Turbulent flow

Figure 12.8 (a) Laminar and (b) turbulent flow.

pipe cross-section. Laminar flow is only possible if the fluid has high viscosity but low flow velocity. If the fluid viscosity is low and flow velocity is high, the flow becomes turbulent.

In **turbulent flow**, eddy currents form, which do not travel in straight paths and collide with one another (Figure 12.8(b)). The conditions needed for transitional flow are in between those for laminar flow and turbulent flow.

Reynolds' number (R_e) may be used to predict the type of flow:

$$R_e = \frac{\rho v D}{\mu}$$

where ρ is the density of the fluid, v the flow velocity, D the pipe diameter and μ the coefficient of absolute viscosity.

$$\text{For laminar flow} : R_e < 2300$$
$$\text{For transitional flow} : 2300 < R_e < 4000$$
$$\text{For turbulent flow} : R_e > 4000$$

12.5.1 Flow Rate

The flow of liquids in pipes is very complex, but in order to derive some important relationships, an ideal liquid is considered one that is non-viscous and incompressible.

Consider a pipe of uniform cross-sectional area, A m^2, containing liquid flowing at a velocity of v m/s.

$$\text{Volume of water flowing per second} \left(\text{or flow rate}\right) = \text{Area} \times \text{length}$$
$$= \text{Area} \times \text{distance travelled per second}$$
$$= \text{Area} \times \text{velocity of flow}$$

Therefore, flow rate, $Q = Av$ (Units: m^3/s.)

This equation can also be applied to pipework consisting of pipes of different diameters, as shown in Figure 12.9. The quantity of liquid flowing through each pipe in a given interval of time must be the same, i.e.:

$$Q = A_1 v_1 = A_2 v_2$$

This equation means that the velocity of the liquid must increase as it enters the narrower pipe to maintain a constant flow rate.

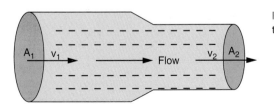

Figure 12.9 Flow through pipework with two cross-sections.

Example 12.3   A pipeline is made by joining two pipes with diameters of 150 mm and 100 mm. If 2.1 m^3 of water flows through the pipe per minute and the pipes run full, find the flow velocity in each pipe.

Solution:
In order to use consistent units, the pipe sizes are converted from millimetres into metres.

Cross-sectional area of 150 mm diameter pipe, $A_1 = \pi \left(r_1 \right)^2$

$$= \pi \left(0.075 \right)^2 \left(r_1 = 75 \, mm, \, or \, 0.075 \, m \right)$$

$$= 0.01767 \, m^2$$

Cross-sectional area of 100 mm diameter pipe, $A_2 = \pi \left(r_2 \right)^2$

$$= \pi \left(0.050 \right)^2 \left(r_2 = 50 \, mm, \, or \, 0.050 \, m \right)$$

$$= 0.007854 \, m^2$$

Flow rate, $Q = 2.1 \, m^3$ per minute

$$= 2.1 \div 60 = 0.035 \, m^3/s$$

Now,

$$Q = A_1 v_1 = A_2 v_2$$

From the above equation,

$$v_1 = \frac{Q}{A_1} = \frac{0.035}{0.01767} = 1.981 \, m/s$$

$$v_2 = \frac{Q}{A_2} = \frac{0.035}{0.007854} = 4.456 \, m/s$$

12.5.2   Bernoulli's Theorem

Bernoulli's theorem is used to determine the energy of a moving liquid. It states that the total energy possessed by a moving liquid is constant, assuming that the loss of energy due to friction and other effects is negligible, and can be expressed as:

Total energy = pressure energy + kinetic energy + potential energy = constant

Pressure Energy
Consider a liquid flowing in a pipe from section 1 to section 2 (see Figure 12.10). If L is the length of the pipe between sections 1 and 2, p is the pressure and A the cross-sectional area of the pipe, then the force exerted by the flowing water will be equal to:

Force exerted = p × A

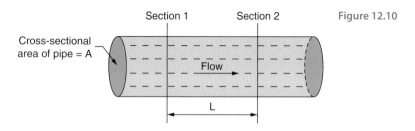

Figure 12.10

Work done in moving the liquid from section 1 to section 2 is:

$$= \text{force} \times \text{distance}$$
$$= p \times A \times L \quad \left(\text{force} = \text{pressure} \times \text{area}\right)$$
$$= p \times V \quad \left(\text{volume}, V = A \times L\right) \tag{12.5}$$

Weight of the liquid, $W = \text{Volume} \times \text{unit weight}$

$$W = V \times w \quad \left(\text{unit weight}, w = \rho \times g\right)$$
$$\text{or}, V = \frac{W}{w} \tag{12.6}$$

From Equations (12.5) and (12.6), pressure energy $= \dfrac{pW}{w}$ (see Section 1.14).

Kinetic Energy

Kinetic energy can be defined as the energy due to motion of the mass of liquid.

$$\text{Kinetic energy} = \frac{1}{2}mv^2$$
$$= \frac{1}{2} \times \frac{W}{g}v^2 \quad \left(\text{mass}, m = W/g\right)$$
$$= \frac{Wv^2}{2g}$$

Potential Energy

If a body of mass m (kg) is raised to a height of z metres, the work done is given by:

$$\text{Work done} = \text{force} \times \text{distance}$$
$$= m \times g \times z$$

If this body is allowed to fall from height z to the original level, it will give out energy that is the same as the work done to raise it.

Therefore, the potential energy of a liquid $= mgz = Wz$.

$$\text{Total energy}, E = \text{pressure energy} + \text{kinetic energy} + \text{potential energy}$$
$$= \frac{pW}{w} + \frac{Wv^2}{2g} + Wz$$

Figure 12.11 Total energy of liquid flow in pipework.

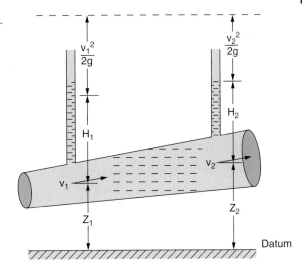

For simplification, we can look at the energy per kilogram of the liquid at any point in the system. This is done by dividing the equation by W.

$$\frac{E}{W} = \frac{p}{w} + \frac{v^2}{2g} + z$$

As stated earlier, energy must be constant; therefore, energy at two points in a pipe (Figure 12.11) is equal. Assuming no friction, we have Bernoulli's equation:

$$\frac{p_1}{w} + \frac{v_1^2}{2g} + z_1 = \frac{p_2}{w} + \frac{v_2^2}{2g} + z_2 = \text{constant}$$

Example 12.4 An inclined pipe decreases uniformly from 150 mm diameter to 100 mm diameter in the direction of water flow. The pressure at the narrow section of the pipe is $2 \times 10^5 \, \text{N/m}^2$, when the flow rate is 0.08 m³/s. The smaller section of the pipe is 0.5 m above the larger section, vertically. Assuming there are no friction losses, calculate the pressure at the 150 mm section. The unit weight of water is 9810 N/m³.

Solution:
In this question we need to take into account the heights of the pipe ends above an arbitrarily fixed datum. This is shown in Figure 12.12.

Figure 12.12

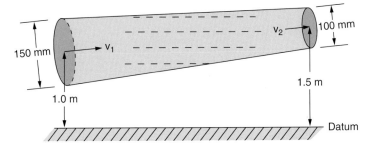

Flow rate, $Q = A_1 v_1 = A_2 v_2$

$$0.08 = \left(\pi \times 0.075^2\right) v_1 = \left(\pi \times 0.05^2\right) v_2$$

$$v_1 = \frac{0.08}{\pi \times 0.075^2} = 4.527 \, \text{m/s}$$

$$v_2 = \frac{0.08}{\pi \times 0.05^2} = 10.186 \, \text{m/s}$$

Pressure at 150 mm section $= p_1$

Pressure at 100 mm section, $p_2 = 2 \times 10^5 \, \text{N/m}^2$

Unit weight (w) of water $= 9810 \, \text{N/m}^3$

Put these values in Bernoulli's equation:

$$\frac{p_1}{w} + \frac{v_1^2}{2g} + z_1 = \frac{p_2}{w} + \frac{v_2^2}{2g} + z_2$$

$$\frac{p_1}{9810} + \frac{4.527^2}{2 \times 9.81} + 1.0 = \frac{2 \times 10^5}{9810} + \frac{10.186^2}{2 \times 9.81} + 1.5$$

$$\frac{p_1}{9810} + 1.045 + 1.0 = 20.387 + 5.288 + 1.5$$

$$\frac{p_1}{9810} = 20.387 + 5.288 + 1.5 - 1.045 - 1.0$$

$$p_1 = 25.13 \times 9810 = 246525.3 \, \text{N/m}^2$$

$$= 246.525 \, \text{kN/m}^2$$

12.5.3 The Venturimeter

The venturimeter is a device used to measure the quantity of a liquid flowing through a pipe. The simplest form, shown in Figure 12.13, consists of two pipes which narrow to form a throat, and is provided with pressure gauges in the pipe and the throat. As a liquid flows through the meter, its velocity increases at the throat due to the reduced cross-section. The kinetic energy at the throat increases, but the pressure energy must reduce to keep the total energy constant.

$$\text{Flow rate, } Q = \frac{KAa\sqrt{2gH}}{\sqrt{\left(A^2 - a^2\right)}}$$

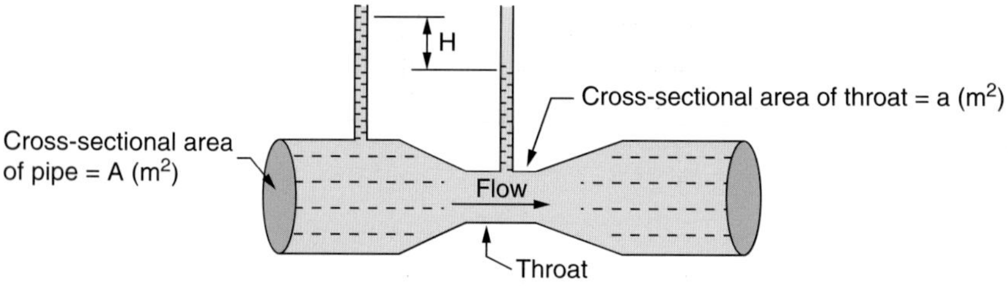

Figure 12.13 Venturimeter.

where A and a are the cross-sectional areas of the pipe and the throat, respectively; H is the difference between the pressure heads; K is the meter coefficient that is introduced to allow for the loss in head due to friction. A typical value of K is 0.97.

Example 12.5 A venturimeter is provided in a 150 mm diameter water pipe. The diameter of the throat is 90 mm and the pressure heads in the main pipe and the throat are 15.0 m and 13.0 m, respectively. Calculate the flow rate in the pipe if the meter coefficient is 0.97 and the acceleration due to gravity is 9.81 ms^{-2}.

Solution:

$$\text{Area of the 150 mm diameter pipe, } A = \pi r^2 \left(r = 0.075\text{ m}\right)$$
$$= \pi(0.075)^2 = 0.01767\text{ m}^2$$

$$\text{Area of the 90 mm diameter throat, } a = \pi r^2 \left(r = 0.04\text{ m}\right)$$
$$= \pi(0.045)^2 = 0.006362\text{ m}^2$$

Difference between the pressure heads, $H = 15.0 - 13.0 = 2.0$ m

$$\text{Flow rate, } Q = \frac{KAa\sqrt{2gH}}{\sqrt{\left(A^2 - a^2\right)}}$$
$$= \frac{0.97 \times 0.01767 \times 0.006362\sqrt{2 \times 9.81 \times 2.0}}{\sqrt{(0.01767)^2 - (0.006362)^2}}$$
$$= 0.0414\text{ m}^3/\text{s}$$

12.5.4 Flow in Pipes: Energy Loss

The Bernoulli theorem assumes that the total energy of a liquid flowing in a pipe remains constant. However, due to the roughness of the pipe, there is a continuous loss of energy due to friction between the pipe surface and the flowing liquid. If a pipe is running full, Darcy's formula can be used to find the loss of pressure head (H):

$$H = \frac{4fLv^2}{2gD}$$

where H = loss of pressure head (m)
  f = friction factor
  L = length of the pipe (m)
  v = mean velocity of flow (m/s)
  D = diameter of the pipe (m)

The friction factor, f, is variable; it depends on the roughness of the pipe material and the Reynolds' number of the flow. For laminar flow, $f = \dfrac{64}{R_e}$; for other types of flow, charts may be used to find the value of f.

Example 12.6   A 200 m long pipe has a diameter of 600 mm. Use Darcy's formula to predict the pressure head loss if the pipe runs full bore with water at a velocity of 2.0 m/s. Assume Darcy's friction factor = 0.01.

Solution:

$$f = 0.01; \ L = 200 \text{ m}; \ v = 2.0 \text{ m/s}; \ D = 0.600 \text{ m}; \ g = 9.81 \text{ m/s}^2$$

$$\text{Loss of pressure head, } H = \frac{4fLv^2}{2gD}$$

$$= \frac{4 \times 0.01 \times 200 \times 2.0^2}{2 \times 9.81 \times 0.600}$$

$$= 2.718 \text{m}$$

12.5.5   Flow in Open Channels

In this section, the Chézy equation and the Manning equation will be discussed and used to calculate the velocity of liquid flow in open channels like drains, partly filled pipes, culverts etc. As the flow in an open channel takes place under atmospheric pressure, it is important to provide the channel with the necessary slope.

Chézy, a French engineer, developed an equation to determine the velocity of water flow in a canal, and considered factors like the slope, the cross-sectional area of the flow and the wetted perimeter (see Figure 12.14). The formula may, however, be applied to any open channel:

$$v = c\sqrt{RS}$$

where v = the average velocity (m/s)
   c = Chézy roughness coefficient (varies from 5 to 77)
   R = hydraulic radius (m)
   S = slope or gradient of the channel

The hydraulic radius of an open channel is the ratio of its cross-sectional area of flow and wetted perimeter:

$$\text{Hydraulic radius, } R = \frac{\text{Cross-sectional area of flow}}{\text{Wetted perimeter}}$$

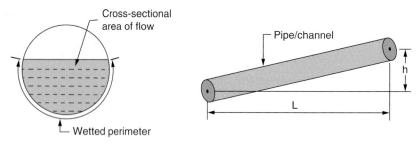

Figure 12.14 Flow in an open channel.

Figure 12.15 Gradient of a channel.

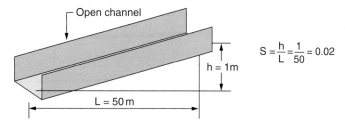

$$S = \frac{h}{L} = \frac{1}{50} = 0.02$$

The gradient of an open channel is usually determined from a short length. For example, the gradient may be given as 1 in 50, which means that for every 50 m horizontal distance (L), the vertical distance between the ends of the channel (h) is 1 m, as shown in Figure 12.15.

Slope, $S = \dfrac{h}{L}$

Robert Manning found that the known data fit the Chézy equation much better if $R^{2/3}$ was used, rather than $R^{1/2}$. The Manning equation is:

$$v = \frac{1}{n} R^{2/3} S^{1/2}$$

where v = the average velocity (m/s)
    n = Manning's roughness coefficient
    R = hydraulic radius (m)
    S = slope or gradient of the channel

Example 12.7 A circular drain of 200 mm diameter runs half-full of water and is laid at a gradient of 1 in 50. If the Chézy coefficient is $50\,\mathrm{m}^{1/2}/\mathrm{s}$, calculate the flow velocity in the drain.

Solution:

$$\text{Cross-sectional area of flow} = \frac{\pi r^2}{2} \quad (r = 0.1\,\mathrm{m})$$
$$= \frac{\pi \times 0.1^2}{2} = 0.0157\,\mathrm{m}^2$$

$$\text{Wetted perimeter} = \frac{2\pi r}{2} = \pi r$$
$$= \pi \times 0.1 = 0.31416\,\mathrm{m}$$

$$\text{Hydraulic radius, R} = \frac{\text{Cross-sectional area of flow}}{\text{Wetted perimeter}}$$
$$= \frac{0.0157}{0.31416} = 0.05\,\mathrm{m}$$

$$\text{Slope, S} = \frac{1}{50} = 0.02$$

$$\text{Velocity, v} = c\sqrt{RS} = 50\sqrt{0.05 \times 0.02}$$
$$= 50 \times 0.0316 = 1.581\,\mathrm{m/s}$$

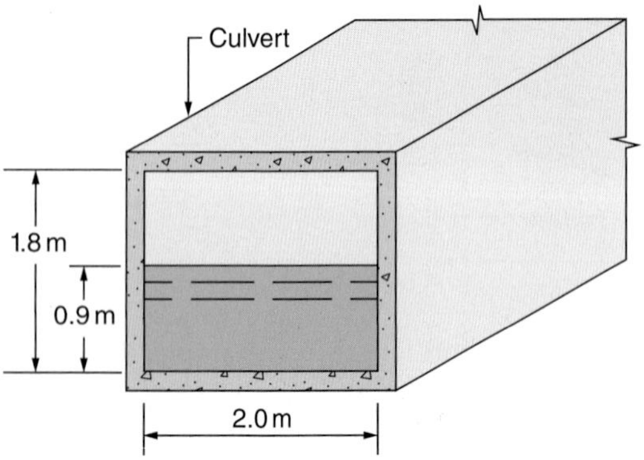

Figure 12.16

Example 12.8 A rectangular culvert, shown in Figure 12.16, runs half-full of water and is laid at a gradient of 1 in 100. Find the velocity of water flow in the culvert using the Chézy equation and the Manning equation, if:

a) The Chézy coefficient is $50\,\mathrm{m}^{1/2}/\mathrm{s}$.
b) Manning's roughness coefficient is 0.012.

Solution:

a) Cross-sectional area of flow $= 2.0 \times 0.9 = 1.80\,\mathrm{m}^2$

Wetted perimeter $= 0.9 + 2.0 + 0.9 = 3.8\,\mathrm{m}$

$$\text{Hydraulic radius, R} = \frac{\text{Cross-sectional area of flow}}{\text{Wetted perimeter}}$$

$$= \frac{1.8}{3.8} = 0.4737\,\mathrm{m}$$

$$\text{Slope, S} = \frac{1}{100} = 0.01$$

$$\begin{aligned}\text{Chézy equation, v} &= c\sqrt{RS} = 50\sqrt{0.4737 \times 0.01}\\ &= 50 \times 0.0688 = 3.441\,\mathrm{m/s}\end{aligned}$$

b) The Manning equation is: $v = \dfrac{1}{n}R^{2/3}S^{1/2}$

$$= \frac{1}{0.012} \times 0.4737^{2/3} \times 0.01^{1/2}$$
$$= 5.064\,\mathrm{m/s}$$

Exercise 12.1

1   A rectangular tank whose base measures 3 m by 2 m contains water to a depth of 1.5 m. Calculate a) the force exerted by the water on the base and the sides; b) the height of the centre of pressure above the base of each side of the tank. Density of water, $\rho = 1000\,\mathrm{kg/m}^3$ and $g = 9.81\,\mathrm{ms}^{-2}$.

2   A concrete dam 300 m long has a vertical face on the reservoir side. Calculate a) the thrust on the dam if the depth of water is 24 m; b) the point of action of this thrust.
Given: $\rho = 1000\,\text{kg/m}^3$ and $g = 9.81\,\text{ms}^{-2}$.

3   A lock gate is 20 m wide. On one side, the water is 6 m above the lower edge of the gate and on the other side the water is 1.5 m above the lower edge of the gate. Calculate:

A The resultant force on the lock gate.
B The point of action of the resultant force.

Given: $\rho = 1000\,\text{kg/m}^3$ and $g = 9.81\,\text{ms}^{-2}$.

4   A water tank measuring 6 m × 6 m × 3 m high has a rectangular door in one of the sides. The door is 1.2 m wide and 1.4 m high, and has a horizontal hinge at the top (see Figure 12.17). If the water is level with the hinge, calculate the horizontal force that must be applied to the lower edge of the door to keep it shut.

(Hint: Determine the thrust on the door and its moment about the hinge. This should be equal to the moment about the hinge due to the applied force.)

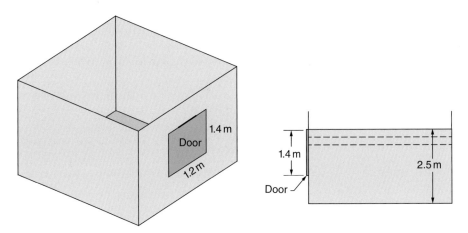

Figure 12.17

5   A pipeline is made by joining two pipes with diameters of 200 mm and 150 mm, respectively. If 2.4 m³ of water flows through the pipe per minute and the pipes run full, find the flow velocity in each pipe.

6   A horizontal pipe decreases uniformly from 150 mm diameter to 100 mm diameter in the direction of the flow of water. The pressure at the narrow section of the pipe is $2.2 \times 10^5\,\text{N/m}^2$, when the flow rate is 0.12 m³/s. Assuming that there are no friction losses, calculate the pressure at the 150 mm section.

7   A venturimeter is provided in a 150 mm diameter water pipe. The diameter of the throat is 75 mm and the pressure heads in the main pipe and the throat are 15.0 m and 12.1 m, respectively. Calculate the flow rate in the pipe if the meter coefficient is 0.97 and the acceleration due to gravity is 9.81 ms⁻².

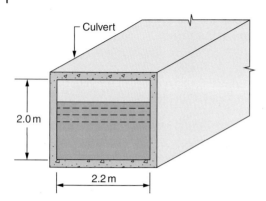

Figure 12.18

8 A rectangular culvert, shown in Figure 12.18, runs three-quarters full of water and is laid at a gradient of 1 in 200. Find the velocity of water flow in the culvert using the Chézy equation and the Manning equation. The Chézy coefficient is $50\,\mathrm{m}^{1/2}/\mathrm{s}$ and Manning's roughness coefficient is 0.011.

References/Further Reading

1 Sherwin, K. and Horsley, M. (1995). *Thermofluids*. Abingdon: Spon Press.
2 Fox, R.W., McDonald, A.T. and Pritchard, P.J. (2009). *Introduction to Fluid Mechanics*. Hoboken: John Wiley & Sons, Inc.

13

Sound

13.1   Introduction

Sound is a physical sensation produced by pressure variations in the air that are caused by sources of vibration. The sound waves, produced by a vibrating object, are transmitted through the medium (solid, liquid or gas) by means of successive compressions and rarefactions of its molecules. Sound waves cannot travel through a vacuum.

Figure 13.1(b) shows a vibrating tuning fork that is sending out sound vibrations due to the outward and inward movements of its prongs. When the prongs move outwards, the molecules of air are brought closer and hence a compression (C) results. This increases the pressure of the air. When the prongs move inwards, the molecules of air move wider apart, or a rarefaction (R) occurs. A rarefaction will create a region with lower pressure. These pressure variations are then passed on to the neighbouring layers of air, which in turn increase and decrease the pressure of further layers. This phenomenon is known as the **sound wave**.

In Figure 13.1(a)/(b), the vertical lines represent the layers of molecules of air. A close spacing of these lines shows higher pressure whereas a wide spacing shows reduced pressure. Although the molecules of air keep on moving to and fro while the tuning fork vibrates, the sound wave (or the sound energy) travels outwards in all directions. The molecules of air come back to their original positions when the vibrations stop.

Construction Science and Materials, Second Edition. Surinder Singh Virdi.
© 2017 John Wiley & Sons Ltd. Published 2017 by John Wiley & Sons Ltd.
Companion website: www.wiley.com/go/virdiconstructionscience2e

(a)

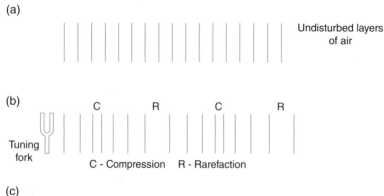

Undisturbed layers
of air

(b)

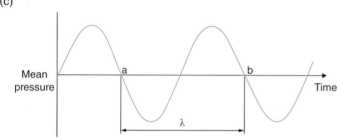

C - Compression   R - Rarefaction

(c)

Figure 13.1

13.2 Frequency, Wavelength and Velocity of Sound

13.2.1 Frequency (f)

The frequency of sound is defined as the number of complete waves, or vibrations, produced per second. Figure 13.1(c), which represents a pure note of single frequency, shows one complete wave from point a to point b. (Unit: hertz (Hz).)

In practice, most sounds are a mixture of frequencies. The audible range for human beings is between 20 Hz and 20000 Hz, however, the upper limit lowers with age.

13.2.2 Wavelength (λ)

The wavelength of sound is defined as the distance that the wave travels in the course of one complete vibration (Figure 13.1(c)). (Unit: metre (m).)

13.2.3 Velocity (v)

If the frequency of a sound is f hertz, then the wave will travel $f \times \lambda$ metres per second.

In other words, the velocity of sound,

$$v = f \times \lambda.$$

(Units: m/s)

A comparison of the velocity of sound through air, water and some solid materials is given in Table 13.1.

Table 13.1 Velocity of sound.

Medium	Velocity (m/s)
Air (0 °C)	331
Air (20 °C)	344
Fresh water (20 °C)	1481
Sea water (20 °C)	1522
Wood	3000–4000
Bricks	4000
Aluminium	5100
Steel	5100
Glass	5000–6000

The velocity of sound in a solid depends on its density and modulus of elasticity. Sound travels at a higher velocity in solids as they have high values of modulus of elasticity. The following relationship may be used to determine the velocity of sound in solid materials:

Velocity of sound in solids, $v = \sqrt{\dfrac{E}{d}}$

where E = modulus of elasticity of the material (N/m^2)
d = density of the material (kg/m^3)

Example 13.1 A vibrating tuning fork produces a sound wave of 256 Hz frequency and 1.30 m wavelength. Calculate the velocity of sound in air.

Solution:

$$\text{Velocity}, v = f \times \lambda$$
$$= 256 \times 1.30 = \mathbf{332.8\,m/s}$$

Example 13.2 A musical instrument vibrates at a frequency of 500 Hz. The time for the transmission of vibrations (in air) through a distance of 50 m is 0.147 seconds. Calculate the velocity and wavelength of the sound vibrations in air.

Solution:

$$\text{Velocity}, v = \frac{\text{Distance}}{\text{Time}}$$
$$= \frac{50}{0.147} = \mathbf{340.14\,m/s}$$

$$\text{Also}, v = f \times \lambda$$
$$340.14 = 500 \times \lambda$$
$$\text{or}, \lambda\ (\text{wavelength}) = \frac{340.14}{500} = \mathbf{0.68\,m}$$

Example 13.3   Find the velocity of sound in steel if the density and the modulus of elasticity of steel are $7850\,\text{kg/m}^3$ and $2\times10^2\,\text{kN/mm}^2$ respectively.

Solution:

$$\text{Velocity of sound, v} = \sqrt{\frac{E}{d}}$$

$E = 2\times10^2\,\text{kN/mm}^2$, but in the above formula the modulus of elasticity should be in N/m^2

$$= 2\times10^2\times\left(1\times10^9\right)\text{N/m}^2\left(1\text{kN/mm}^2 = 1\times10^9\,\text{N/m}^2\right)$$

$$= 2\times10^{11}\,\text{N/m}^2$$

$$d = 7850\,\text{kg/m}^3$$

$$v = \sqrt{\frac{E}{d}} = \sqrt{\frac{2\times10^{11}}{7850}}$$

$$= \mathbf{5047.5\,m/s}$$

13.3   Measurement of Sound

Sound is produced when a vibrating source disturbs the air and transmits a sound wave through it in all directions. Our ears are not sensitive to the total sound energy but are sensitive to the rate at which the energy reaches them. This rate determines the loudness of a sound source; therefore, the **power** of a source of sound is the rate at which the energy is produced. The unit of sound power is the watt (W).

Sound waves travel in all directions away from the source. The surface area on which the waves fall increases as the distance from the sound source increases. Hence, the intensity of the sound waves decreases as we move away from the source. **Sound intensity** is defined as the sound power per unit area over which the sound waves fall.

$$\text{Sound intensity, I} = \frac{\text{Sound power}\left(\text{watts}\right)}{\text{Surface area}\left(\text{m}^2\right)}$$

$$\left(\text{Units: W/m}^2\right)$$

Since audible sound consists of pressure waves, sound pressure may also be used to quantify sound. **Sound pressure** (p) is defined as the pressure variations relative to atmospheric pressure caused by the sound wave. (Unit: N/m^2 or pascals (Pa).)

$$1\,\text{N/m}^2 = 1\,\text{Pa}$$

Sound intensity is proportional to the square of the sound pressure and is expressed as:

$$I = \frac{p^2}{\rho v}$$

where I = sound intensity in W/m^2
p = sound pressure in N/m^2
ρ = density of the medium (density of air at $20\,^\circ\text{C} = 1.2\,\text{kg/m}^3$)
v = velocity of sound

13.3.1   Threshold Values of Sound

Many sound measurements are made relative to a standard **threshold of hearing**, which is defined as the weakest sound that the average human ear can detect.

The threshold of hearing varies with frequency, but at 1000 hertz the values are:

$$\text{In terms of intensity}: I_0 = 1 \times 10^{-12}\,\text{W/m}^2$$
$$\text{In terms of pressure}: p_0 = 2 \times 10^{-5}\,\text{N/m}^2 \left(\text{or Pa}\right)$$

The strongest sound that the human ear can tolerate is defined as the **threshold of pain**. Generally, the threshold of pain decreases with age, but the damage done by loud sounds to human hearing is the same, irrespective of age. The values of the threshold of pain are:

$$\text{In terms of intensity}: I_p = 100\,\text{W/m}^2$$
$$\text{In terms of pressure}: p_p = 200\,\text{N/m}^2 \left(\text{or Pa}\right)$$

13.3.2   The Decibel Scale

The strength of sound may be specified in terms of sound intensity or sound pressure. However, the range of values between the threshold of hearing and the threshold of pain is so large that their use is not convenient. Another problem is that the response of the ear to any change in pressure is dependent on the existing pressure, i.e. it is a comparison between the new and the old sound levels. In order to have a convenient range of values that compare reasonably well with the way that the ear responds to sound, the **bel** scale was devised. The bel scale is a logarithmic scale and is basically a ratio of two sound levels:

$$\text{Sound intensity level}\left(\text{in bel}\right) = \log_{10}\left[\frac{I_1}{I_0}\right]$$

where I_1 and I_0 are the sound intensities to be compared.

I_0 is the reference-level sound intensity; in most cases it is the threshold of hearing, i.e. 1×10^{-12} W/m².
A tenfold increase in intensity of the sound equates to an increase of one unit on the bel scale. For example, if a sound intensity increases ten times from 0.1 W/m² to 1.0 W/m², the corresponding increase on the bel scale is:

$$\text{Sound level} = \log_{10}\left[\frac{I_1}{I_0}\right] = \log_{10}\left[\frac{1.0}{0.1}\right] = \log_{10}10 = 1.0\,\text{bel}$$

The **decibel** (dB) value of the intensity of a sound is ten times the bel value:

$$\text{Sound intensity level}\left(\text{in dB}\right) = 10 \times \text{bel value} = 10\log_{10}\left[\frac{I_1}{I_0}\right]$$

$$\text{Also, sound pressure level}\left(\text{in dB}\right) = 20\log_{10}\left[\frac{p_1}{p_0}\right]$$

where p_1 = sound pressure level in N/m²
p_0 = reference value of sound pressure. In most cases, it is the threshold of hearing, i.e. 2×10^{-5} N/m²

Table 13.2

Sound pressure	Intensity of sound	Sound level		Sound source/effect
		Bel	Decibels	
2×10^{-5}	1×10^{-12}	0	0	Threshold of hearing
	1×10^{-11}	1	10	Rustling leaves
2×10^{-4}	1×10^{-10}	2	20	Whisper
	1×10^{-9}	3	30	Quiet library
2×10^{-3}	1×10^{-8}	4	40	Quiet office
	1×10^{-7}	5	50	Average office
2×10^{-2}	1×10^{-6}	6	60	Normal conversation
	1×10^{-5}	7	70	Noisy restaurant
2×10^{-1}	1×10^{-4}	8	80	Normal traffic
	1×10^{-3}	9	90	Heavy traffic
2×10^{0}	1×10^{-2}	10	100	Near a pneumatic drill
	1×10^{-1}	11	110	Riveting machine
2×10^{1}	1×10^{0}	12	120	Chain saw
	1×10^{1}	13	130	Aircraft take-off (threshold of pain)
2×10^{2}	1×10^{2}	14	140	Intense physical pain

Table 13.2 shows the sound levels between the thresholds of hearing and pain, and their equivalent values in bel and decibels.

Example 13.4   The intensity of a sound is 2.5×10^{-6} W/m². If the threshold of hearing is 1×10^{-12} W/m², find the sound intensity level of this sound in decibels.

Solution:

$$\text{Sound intensity level} = 10\log_{10}\left[\frac{I_1}{I_0}\right]$$

$$I_1 = 2.5 \times 10^{-6}; \quad I_0 = 1 \times 10^{-12}$$

$$\text{Sound intensity level} = 10\log_{10}\left[\frac{I_1}{I_0}\right]$$

$$= 10\log_{10}\left[\frac{2.5 \times 10^{-6}}{1 \times 10^{-12}}\right]$$

$$= 10\log_{10}\left[2.5 \times 10^{6}\right]$$

$$= 10 \times 6.398 = \mathbf{63.98\,dB}$$

Example 13.5 Find the increase in sound level if the sound intensity is:

a) doubled;
b) trebled.

Solution:

a) In this question, I_1 is twice I_0, therefore, $\dfrac{I_1}{I_0} = 2$

$$\text{Sound intensity level} = 10\log_{10}\left[\frac{I_1}{I_0}\right]$$
$$= 10\log_{10} 2 = \mathbf{3.01\,dB}$$

(When the sound intensity is doubled, the increase in the number of decibels is 3.01.)

b) In this question, I_1 is three times I_0, therefore, $\dfrac{I_1}{I_0} = 3$

$$\text{Sound intensity level} = 10\log_{10}\left[\frac{I_1}{I_0}\right]$$
$$= 10\log_{10} 3 = \mathbf{4.77\,dB}$$

13.4   Addition of Sound Levels

There are many instances when two or more sources produce sound simultaneously and it is important to find the overall effect. The individual sound levels, which may be specified either as sound intensity (W/m^2) or as sound pressure (N/m^2), may be added like ordinary numbers to find the resulting effect. However, if the sound levels are given in decibels, they cannot be added like ordinary numbers. Since the decibel scale is a logarithmic scale, the sound levels in decibels are converted to sound intensities, or to sound pressures, and then added. The addition of sound levels can be done by using either the equations given in Section 13.3.2 or a scale explained in Section 13.4.1.

13.4.1   Approximate Addition of Sound Levels

The decibel levels can be added quickly by using the scale shown in Figure 13.2.

The calculation first involves finding the difference between the two sound levels. The difference is located on the upper scale, and the correction determined from the lower scale. The correction is then added to the higher sound level to find the overall sound level.

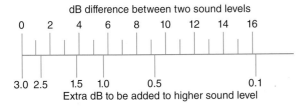

Figure 13.2 Addition of decibel levels.

Example 13.6   Two car engines were started at the same time and the corresponding noise levels measured at a certain point were 5×10^{-1} Pa and 1.6 Pa. Calculate the overall noise level in decibels.

Solution:

Overall noise level, $p_1 = 5 \times 10^{-1} + 1.6 \, \text{Pa} = 2.1 \, \text{Pa}$

$$\text{Overall noise level (in dB)} = 20 \log_{10} \left[\frac{p_1}{p_0} \right]$$

$$= 20 \log_{10} \left[\frac{2.1}{2 \times 10^{-5}} \right]$$

$$= 20 \log_{10} \left(1.05 \times 10^5 \right)$$

$$= \mathbf{100.42 \, dB}$$

Example 13.7   Three machines working at the same time produce noise levels of 70 dB, 75 dB and 82 dB. Find the overall noise level using the approximate, and the analytical, method.

Solution:
As explained in Section 13.4, the decibel levels of noise cannot be added like ordinary numbers; $70 + 75 + 82 = 227$ dB will be totally wrong. The overall noise level can be determined by two methods: the approximate and the analytical, method, which are given below.

a) The approximate method
    In this method, the scale shown in Figure 13.2 is used. As only two noise levels can be dealt with at a time, consider 70 dB and 75 dB first.

    Difference $= 75 - 70 = 5 \, \text{dB}$

    The difference of 5 dB is located on the upper scale, and the correction found from the lower scale. Correction $= 1.2 \, \text{dB}$
    Overall noise level of 70 and 75 dB $= 75 + 1.2 = 76.2 \, \text{dB}$
    Now consider 76.2 and 82 dB:

    difference $= 82 - 76.2 = 5.8 \, \text{dB}$

    correction $= 1.0 \, \text{dB}$

    Overall noise level $= 82 + 1.0 = \mathbf{83 \, dB}$

b) The analytical method
    In this method, we need to change each decibel level into sound intensity. The sound intensities are added and the resulting noise level converted back into decibels. Let us first convert 70 dB into sound intensity.

    From Section 13.3.2, sound intensity level $(\text{dB}) = 10 \log_{10} \left[\frac{I_1}{I_0} \right]$

    $$70 = 10 \log_{10} \left[\frac{I_1}{I_0} \right]$$

    $$\frac{70}{10} = \log_{10} \left[\frac{I_1}{I_0} \right] \quad \text{or,} \quad \log_{10} \left[\frac{I_1}{I_0} \right] = 7$$

To cancel out $\log_{10}$ take the antilog of both sides:

$$\text{antilog}\left(\log_{10}\frac{I_1}{I_0}\right) = \text{antilog}\,7$$

$$\frac{I_1}{I_0} = 10^7 \text{ or, } 10000000$$

Similarly, $10\log_{10}\left[\dfrac{I_2}{I_0}\right] = 75$

Take the antilog of both sides:

$$\text{antilog}\left(\log_{10}\frac{I_2}{I_0}\right) = \text{antilog}\,7.5$$

$$\frac{I_2}{I_0} = 3.162 \times 10^7$$

and, $10\log_{10}\left[\dfrac{I_3}{I_0}\right] = 82$

After taking the antilog of both sides, $\dfrac{I_3}{I_0} = 1.585 \times 10^8$

Now we can add all the sound intensities:

$$\frac{I_4}{I_0} = \frac{I_1}{I_0} + \frac{I_2}{I_0} + \frac{I_3}{I_0}$$

$$\frac{I_4}{I_0} = 10^7 + 3.162 \times 10^7 + 1.585 \times 10^8$$

$$= 2.001 \times 10^8$$

Finally, convert the total intensity into decibels:

$$\text{sound intensity level (dB)} = 10\log_{10}\left[\frac{I_4}{I_0}\right]$$

$$= 10\log_{10}\left(2.001 \times 10^8\right)$$

$$= \mathbf{83.01\,dB}$$

13.5 Transmission of Sound in Buildings

Sound is transmitted into buildings as **airborne** sound and **impact** sound. Airborne sound is defined as the sound that is produced by a vibrating object and travels through air as waves. Typical examples are the sound produced by musical instruments, speech, traffic noise etc. Impact sound, also known as structure-borne sound, occurs when a building component is set into vibration by impact, for example, footsteps, vibrating machinery and slammed doors (Figure 13.3).

A building should provide adequate insulation from airborne as well as from impact sound. Sound waves produced by sources like traffic travel through air as airborne sound and fall on the roof, windows, external doors and external walls, and set these elements into vibration. Some of the sound

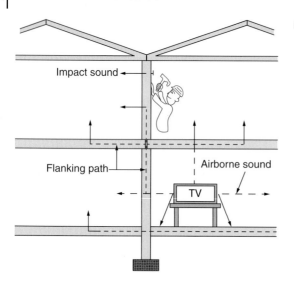

Figure 13.3

energy is used in this process; the rest is transmitted to the interior of the building. Any gaps or cracks which may exist around doors and windows allow an easy passage for the sound waves.

Sound may also travel through walls, floors and other components to other parts of the building by **flanking transmission**, as shown in Figure 13.3.

13.5.1 Noise

Disturbance by noise is one of the most annoying aspects of life in the modern world. There are several sources of noise, for example, noisy neighbours, road traffic, low-flying aircraft etc. Even music is noise when someone is trying to sleep or concentrate on some activity.

Any unwanted sound is called noise. Traffic noise accounts for a major proportion of the total noise generated outside dwellings in the UK. It is caused by the engine noise and the vibrations of tyres on the road surface. Low speeds can reduce noise levels; a 20–24 kph (kilometres per hour) increase in speed results in noise levels rising by 4–5 decibels. Traffic-calming measures may reduce the vehicles' speed but could cause an increase in noise levels if the result is several stop–go vehicle movements.

Aircraft noise also harms people's health. Many people who live under the flight path suffer high levels of annoyance, stress and sleep deprivation.

Exposure to intense noise, for a few hours, causes a temporary loss of hearing sensitivity known as a **temporary threshold shift** (TTS). After some rest, the person usually recovers. If someone is exposed to intense occupational noise over a period of several years, the TTS changes to a permanent threshold shift as there is not enough time to recover.

13.5.2 Requirements of Sound Insulation

Before describing the techniques for achieving sound insulation, it is important to understand the underlying principles of good sound insulation; these are outlined in the following subsections.

Mass
The insulation provided by a partition is directly proportional to its mass. As the mass of a material also depends on its density, dense materials provide greater mass and hence higher sound insulation. Sound waves can easily set light partitions into vibration and transmit noise into a building. Sound

insulation of walls may therefore be improved either by increasing the thickness or by using materials of higher density. However, for external walls, lighter materials are preferred as they provide better thermal insulation.

Generally, doubling the mass of a partition increases its sound insulation by 5 dB. A one-brick-thick wall (215 mm + plaster) has a mass of $415 \, kg/m^2$ and gives an average sound insulation of about 50 dB.

Completeness

The external envelope of a building contains doors for access and windows for light and ventilation. The insulation provided by doors and windows is less than that provided by a brick wall due to lower mass per unit area. The doors and windows are fitted into openings in the wall and it is possible that the gaps between their frames and the wall are not sealed properly. Sound waves can easily pass through gaps and cracks and reduce the sound insulation of an otherwise well-insulated wall. Before other measures are taken to increase the sound insulation of a wall, the elimination of gaps around doors and windows should be considered first.

Discontinuous Construction

Discontinuous construction may be used where we have to use materials that are light in weight, for example cavity wall construction involving aerated concrete blocks. For better results, the cavity should be wider than 50 mm, the gap filled with sound-absorbent materials and the panels not connected by ties or other methods of construction. Typical examples of discontinuous construction are double-glazed windows, cavity walls and upper-floor construction.

Resonance and Coincidence Effects

If the sound waves have the same frequency as the natural frequency of the partition, **resonance** will occur, i.e. the vibrations will be amplified. Resonance occurs at low frequencies and reduces the sound insulation of the building component.

Sound waves can cause a partition to bend inwards and outwards. When the incident sound reaches a partition at angles less than 90°, its projected wavelength may be the same as that of the resonant bending waves. The frequency at which the two wavelengths coincide is known as the **critical frequency**. The phenomenon which occurs at high frequencies is known as **coincidence** and causes efficient transfer of sound energy from one side of the panel to the other. The vibrations are amplified, causing a reduction in the sound insulation.

13.5.3 Sound-Insulation Techniques

Buildings should be designed to reduce the external and internal noise to acceptable levels. Typical examples of **external noise** are traffic, aircraft, trains, construction work and noisy neighbours. The sources of **internal noise** are usually people and their activities, and noisy machines/equipment. They could be either in another part of the building or in the same room as the affected people.

As the sound intensity is inversely proportional to the square of the distance from the source, the building should be sited as far away from the source of noise as possible. This can only be done at the design stage, but may not be feasible due to reasons such as a shortage of suitable building sites.

Noise can be controlled in three ways: at the source; in its path (airborne sound); and at the receiver.

At the Source

Some construction activities and machines in factories create noise that can be controlled at the source. Piling is a construction activity that creates a lot of noise if piling hammers are used. If possible,

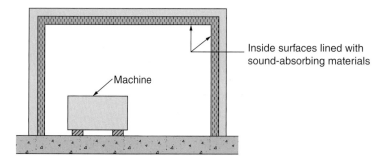

Figure 13.4

alternative methods of piling should be used, but otherwise the top of the pile should be provided with a hardwood or plastic block, called a dolly, to absorb the impact of the pile hammer and hence reduce noise. Machines in factories should be serviced regularly, but if a machine still produces excessive noise, then isolated rooms should be used to enclose such a machine (Figure 13.4). The material used in constructing an isolated room should have an appropriate sound-reduction index (see Section 13.5.5). Where pipes and ducts are used, they should enter through the walls with all the gaps properly sealed with mastic or similar sealant. Any gaps around doors and windows should also be sealed. For the safety of the machine operators or other personnel inside the enclosure, it is important to reduce the noise level within the room. This can be achieved by lining the walls and ceiling with absorbent material like mineral wool. The vibrations produced by a machine can travel through the floor to other parts of the building, and should be minimised by installing the machine on anti-vibration mountings made from springs, rubber, corkboard or other suitable materials.

The most common sources of impact noise are footsteps or objects falling on floors with a hard surface. The impact can be reduced by covering the floor with a thick carpet and underlay or with tiles made of resilient materials. A floating floor may also be used as insulation against impact noise, and involves the provision of a layer of resilient material in between the top surface and the main supporting structure. Figure 13.5 shows two examples of floor construction to achieve sound insulation.

Airborne Sound

Noise can be minimised by putting a barrier in the path of the sound waves. Noise barriers can be made from a variety of materials such as timber, concrete, steel, bricks and plastic. For a barrier to be effective, it must be as dense as possible, of sufficient height and have a mass of about $10\,\text{kg/m}^2$. The cheapest solution is a close-boarded timber fence, high enough to maximise the path difference, as shown in Figure 13.6. Barriers formed by landscaping using earth embankments are quite effective in reflecting and absorbing traffic noise.

Trees and hedges may be used as noise barriers, but they can only be effective if there are several rows of them to provide dense vegetation. It is also important to bear in mind that the rustle of leaves may produce noise levels as high as 40 dB.

Noise from one room to another in the same building can be transmitted as direct sound and/or flanking transmission. Noise as direct sound is transmitted in the same way as noise from external sources and it can be dealt with by providing walls as heavy as possible. Doors and windows are weak points and, if not fitted carefully, they can reduce the overall insulation of a wall. They should fit properly without creating any gap between their frames and the wall. Noise may also travel through the flanking walls, floors and ceilings, which could be reduced by providing a layer of resilient material between the joints of the building elements.

(a)

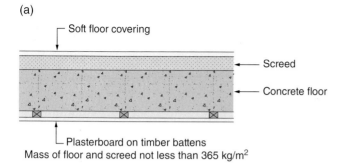

Soft floor covering

Screed

Concrete floor

Plasterboard on timber battens

Mass of floor and screed not less than 365 kg/m²

(b)

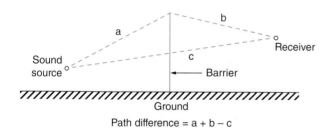

Timber/chipboard flooring

25 mm thick layer of resilient material

Concrete floor minimum mass = 300 kg/m²

Plasterboard on timber battens

Figure 13.5

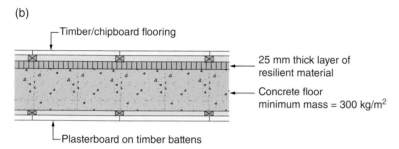

a

b

Sound source

c

Receiver

Barrier

Ground

Path difference = a + b − c

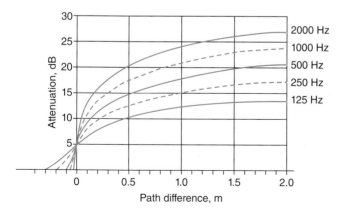

Figure 13.6

At the Receiver

When exposure to intense noise is unavoidable and the methods described previously are not practical, then measures must be taken to protect the receiver. This solution should be considered as a last resort in some situations. In others, this may be the only solution, for example, operatives using pneumatic drills. There are a variety of possibilities, for example, the use of ear protection.

13.5.4 Noise in a Workplace

The Noise Regulations require the employer to take specific action at certain noise levels:

- Lower exposure action values:
    daily or weekly exposure of 80 dB
    peak sound pressure of 135 dB
- Upper exposure action values:
    daily or weekly exposure of 85 dB
    peak sound pressure of 137 dB

There are also levels of noise exposure that must not be exceeded:

- Daily or weekly exposure of 87 dB;
- Peak sound pressure of 140 dB.

At and above a lower exposure action value, suitable hearing protection must be made available to any employee who requests it.

At and above an upper exposure action value, the exposure must be reduced to as low a level as possible by organisational and technical measures, excluding hearing protectors. These measures could include the choice of appropriate equipment, design and layout of workplaces, regular maintenance of equipment and appropriate work schedules. For more details, refer to the Health and Safety Executive's publication, *Noise at Work*.

13.5.5 Measurement of Sound Insulation

The **sound reduction index** (R) of a building element is a measure of the sound insulation (in decibels) that the element provides. Its value depends on the magnitudes of the incident and transmitted sound energy, and is given by the expression:

$$R = 10\log_{10}\left(\frac{\text{Incident sound energy}}{\text{Transmitted sound energy}}\right)$$

Example 13.8 The intensity of traffic noise incident on a building wall is 5×10^{-5} W/m^2. If the wall transmits 2×10^{-7} W/m^2 into the building, find the sound reduction index of the wall.

Solution:

$$\text{Sound reduction index}\left(R\right) = 10\log_{10}\left(\frac{\text{Incident sound energy}}{\text{Transmitted sound energy}}\right)$$

$$= 10\log_{10}\left(\frac{5 \times 10^{-5}}{2 \times 10^{-7}}\right)$$

$$= 10 \times 2.398 = \mathbf{23.98\,dB}$$

Example 13.9 The sound reduction index of a wall is 30 dB. Calculate what percentage of incident sound would be transmitted into the building.

Solution:
Let the incident sound be 100% and the transmitted sound be x%.

$$\text{Sound reduction index }(R) = 10\log_{10}\left(\frac{\text{Incident sound energy}}{\text{Transmitted sound energy}}\right)$$

$$30 = 10\log_{10}\left(\frac{100}{x}\right)$$

$$\frac{30}{10} = \log_{10}\left(\frac{100}{x}\right)$$

Taking the antilog, $\text{antilog } 3 = \text{antilog}\left[\log_{10}\left(\frac{100}{x}\right)\right]$

$$1000 = \frac{100}{x} \quad \text{or, } x = \textbf{0.1}$$

The transmitted sound is **0.1%** of the incident sound.

13.6 Sound Absorption

In this section, the quality of sound produced in a room or a hall will be discussed. When sound is produced in a room, the waves strike the walls, ceiling, floor and other surfaces. Depending on the surface material, some of the sound is reflected, some absorbed and the rest transmitted through the structure, as shown in Figure 13.7.

Different materials and surface finishes absorb different amounts of incident sound and it is the total amount of sound absorbed by the surfaces of a room that affects the quality of sound. Sound absorption coefficients are available for a range of building materials and surface finishes, and are given by the ratio:

$$\text{Sound absorption coefficient }(\alpha) = \frac{\text{Amount of sound energy absorbed by a surface}}{\text{Amount of incident sound energy}}$$

Figure 13.7

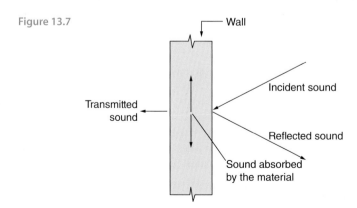

Table 13.3 Absorption coefficients.

Material	Absorption coefficient		
	500 Hz	2000 Hz	4000 Hz
Brickwork	0.03	0.04	0.05
Carpet (medium) on timber floor	0.30	0.50	0.60
Carpet (medium) on concrete floor	0.20	0.30	0.40
Concrete	0.02	0.02	0.02
Curtains (medium fabrics) folded away from walls	0.45	0.50	0.50
Curtains (medium fabrics) straight, close to wall	0.25	0.30	0.40
Doors – panelled	0.08	0.10	0.10
Floor tiles – hard	0.03	0.05	0.05
Glass – 4 mm thick	0.10	0.05	0.05
Linoleum on solid backing	0.05	0.10	0.10
Plaster (gypsum) on a solid backing	0.02	0.04	0.03
Plasterboard ceiling (9.5 mm thick)	0.05	0.07	0.09
Suspended ceiling of plasterboard (large airspace behind)	0.10	0.05	0.05
Wood blocks on solid floor	0.05	0.10	0.10
Window (open)	1.00	1.00	1.00
Other items			
Air (per m^3)	–	0.007	0.02
Audience in fully upholstered seats (per person)	0.46	0.51	0.47
Seats (unoccupied) fully upholstered (per seat)	0.28	0.28	0.35

The best sound-absorber is an open window, as none of the incident sound can be reflected back into the room. Table 13.3 shows typical values of sound absorption coefficients for some common materials.

The absorption of sound by a surface is the product of its area (m^2) and its absorption coefficient. In a room, there are several types of surfaces, for example, carpet as floor finish, plaster on brick/block walls, plasterboard ceiling and furniture. As people absorb sound as well, the total sound-absorption units for a room are the sum of absorption by all the surfaces, including people.

$$\text{Total sound absorption of a room} = \Sigma \left(\text{area of a surface} \times \text{absorption coefficient} \right)$$

The acoustic requirements for a room depend on the purpose for which it is to be used. Some of the requirements for good room acoustics are:

1) Even distribution and adequate levels of sound to all listeners.
2) Adequate insulation against external noise.
3) Appropriate reverberation time.
4) The absence of acoustic defects like echoes, acoustic shadow etc.

Like reverberations, **echoes** are also the result of reflection of sound. The former occur when a reflected sound wave reaches the ear in less than 0.1 second after the original sound wave. If a reflected sound wave reaches the ear in more than 0.1 second after the original sound wave, it is known as an echo.

Figure 13.8 Raking of rows of seats to give an unobstructed path for direct sound.

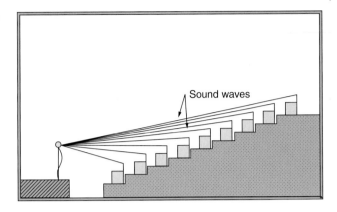

Sound waves

Acoustic shadow is also an acoustic defect. Areas that do not receive any sound are affected by acoustic shadow. In a hall, the front rows receive the direct sound whereas people sitting at the back receive only reflected sound waves. Acoustic shadows also occur behind pillars and under low, deep balconies. To improve the hearing conditions in cinema halls and auditoria, the rows of seats are raked so that the direct sound is not obstructed (see Figure 13.8).

13.6.1 Reverberation

Reverberation is the continued presence of sound in an enclosure due to repeated reflections of sound from the surfaces. Consider a source of sound, for example a CD player, in a room where there is a listener 8.25 m away from the CD player. Sound produced by the CD player may reach the listener directly by path **d** and after reflections from the walls and other surfaces, as shown in Figure 13.9.

$$\text{Time taken for the direct sound to reach the listener} = \frac{\text{Distance}}{\text{Velocity}} = \frac{8.25}{344} = 0.024\,\text{s}$$

$$\text{Time taken for reflected sound } a_1 \text{ to reach the listener} = \frac{\text{Distance}}{\text{Velocity}} = \frac{16.0}{344} = 0.047\,\text{s}$$

Similarly, the times taken for reflected sounds a_2, b_1 and b_2 are 0.038 s, 0.041 s and 0.047 s, respectively. The multiple reflections cause two effects:

1) Distortion due to sound reaching the listener at different times.
2) An increase in the loudness of sound when several waves reach the listener simultaneously. Fortunately, when a sound wave strikes a surface, only a part of it is reflected, the rest is absorbed and/or transmitted.

13.6.2 Reverberation Time

The reverberant sound in a room dies away with time as the sound energy is absorbed by the room surfaces. The time taken for a sound to decay by 60 dB is known as the **reverberation time** (RT). Figure 13.10 shows that a source that produced a sound level of 90 dB was stopped suddenly. The time taken for the sound level to decay by 60 dB (from 90 to 30 dB) is 1.2 s, which is the reverberation time of the room. The main factors that affect the reverberation time of a room are its size and the sound-absorbing properties of its surfaces.

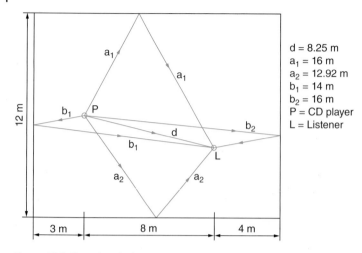

$d = 8.25$ m
$a_1 = 16$ m
$a_2 = 12.92$ m
$b_1 = 14$ m
$b_2 = 16$ m
$P = $ CD player
$L = $ Listener

Figure 13.9 Sound paths in a room.

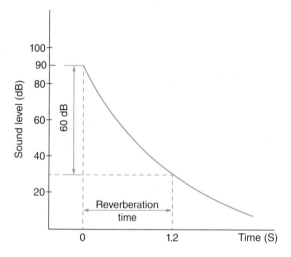

Figure 13.10

The actual reverberation time in a room or a hall may be calculated from Sabine's formula:

$$RT = \frac{0.16V}{A}$$

where RT = reverberation time in seconds (s)
V = volume of room/hall in cubic metres (m³)
A = total absorption of room surfaces (m² sabins)

The optimum reverberation time is different from the actual reverberation time as it depends only on the size of the room and its use. Stephen and Bate's formula may be used to determine the optimum reverberation time:

$$RT\,(\text{optimum}) = r\left[0.0118\,\sqrt[3]{V} + 0.107\right]$$

where RT = reverberation time in seconds (s)

$\qquad$ V = volume of room/hall in cubic metres (m^3)

$\qquad$ r = a constant: 4 for speech; 5 for orchestra; 6 for chorus

Example 13.10 A hall measuring 30 m × 25 m × 6 m high has the following surface finishes:

$\qquad$ floor: medium-pile carpet on solid floor

$\qquad$ walls: plaster on blockwork

$\qquad$ ceiling: plasterboard on joists

$\qquad$ doors: three − 2 m × 2 m high

$\qquad$ windows: ten − 1.5 m × 2 m high

a) Calculate the actual reverberation times at 500 Hz and 2000 Hz if the capacity of the hall is 500 people. Assume that there is an acoustic shadow on 50% of the floor area.
b) Calculate the optimum reverberation time if the hall is to be used for speech.
c) Calculate the optimum reverberation time if the hall is to be used for an orchestra.

Solution:

a) To keep the calculations simple, the area of light fittings will be ignored. Due to the acoustic shadow, 50% of the floor area will be considered in the calculations.

$$\text{Floor area} = \text{ceiling area} = 30 \times 25 = 750 \text{ m}^2$$

$$\text{Wall area} = 2(30 \times 6 + 25 \times 6) = 660 \text{ m}^2$$

$$\text{Area of doors} = 2 \times 2 \times 3 \text{ no.} = 12 \text{ m}^2$$

$$\text{Area of windows} = 1.5 \times 2 \times 10 \text{ no.} = 30 \text{ m}^2$$

$$\text{Net wall area} = 660 - 12 - 30 = 618 \text{ m}^2$$

$$\text{Volume of hall, V} = 30 \times 25 \times 6 = 4500 \text{ m}^3$$

Surface		500 Hz		2000 Hz	
	Area (m^2)	Absorption coefficient	Absorption units	Absorption coefficient	Absorption units
	(a)	(b)	(= a × b)	(c)	(= a × c)
Walls	618	0.02	12.36	0.04	24.72
Floor	50% of 750 = 375	0.20	75.0	0.30	112.5
Ceiling	750	0.05	37.5	0.07	52.5
Windows	30	0.10	3.0	0.05	1.5
Doors	12	0.08	0.96	0.10	1.2
Audience	500 people	0.46	230	0.51	255
Air	Volume = 4500 m^3	–	–	0.007	31.5
			Total (A) = 358.82		Total (A) = 478.92

$$\text{Actual RT at } 500\,\text{Hz} = \frac{0.16V}{A}$$

$$= \frac{0.16 \times 4500}{358.82} = 2.0\,\text{s}$$

$$\text{Actual RT at } 2000\,\text{Hz} = \frac{0.16 \times 4500}{478.92} = 1.50\,\text{s}$$

b) Optimum RT $= r\left[0.0118\sqrt[3]{V} + 0.107\right]$

r (for speech) $= 4$, Volume, $V = 4500\,\text{m}^3$

$$\text{Optimum RT} = 4\left[0.0118 \times \sqrt[3]{4500} + 0.107\right]\left(\sqrt[3]{4500} = \text{cube root of } 4500\right)$$

$$= 4\left[0.0118 \times 16.51 + 0.107\right]$$

$$= 4\left[0.1948 + 0.107\right] = 1.21\,\text{s}$$

The reverberation time at 2000 Hz is slightly more than the optimal value, but at 500 Hz, the actual RT is 0.79 s higher than the optimum. This needs to be brought down to about 1.2 s by using alternative finishes.

c) r (for orchestra) $= 5$, Volume, $V = 4500\,\text{m}^3$

$$\text{Optimum RT} = 5\left[0.0118 \times \sqrt[3]{4500} + 0.107\right]$$

$$= 5\left[0.0118 \times 16.51 + 0.107\right]$$

$$= 1.51\,\text{s}$$

This is very close to the actual reverberation time at 2000 Hz, but 0.49 s lower than the actual reverberation time at 500 Hz. Some of the finishes need to be altered to achieve a match between the actual and the optimum reverberation times.
(A spreadsheet of this example is given in Appendix 2.)

Example 13.11   In Example 13.10 the actual reverberation time at 500 Hz is 2.0 s. Suggest suitable materials to bring the reverberation time down to 1.2 s.

Solution:
Total absorption units (from Example 13.10) = 358.82
    Total absorption units required for a reverberation time of 1.2 s

$$= \frac{0.16V}{1.2} = \frac{0.16 \times 4500}{1.2} = 600$$

Extra absorption units required = 600 − 358.82 = 241.18
Acoustic panels ($\alpha = 0.8$) may be used to provide the extra absorption:

$$\text{Area of acoustic panels required} = \frac{241.18}{0.8} = 301.5\,\text{m}^2$$

13.6.3   Types of Sound Absorbers

The frequency of sound affects the choice of materials that can be used as sound absorbers. Usually, absorption is considered for low-frequency, medium-frequency and high-frequency sounds. The absorption of low-frequency sound (up to about 400 Hz) is achieved by panels of flexible material provided over a solid background with an air gap in between. For slightly better results, the air gap may be filled with porous materials like mineral wool or glass fibre, as shown in Figure 13.11.

A cavity absorber, or Helmholtz resonator, which can be used to absorb sound of a single specific frequency, consists of an enclosure of air with one narrow opening (Figure 13.12). Sound enters the air enclosure, resonates and loses energy. A simple solution is to provide a perforated panel on a wall or a ceiling with an airspace in between. The absorption of sound can be enhanced by filling the airspace with mineral wool or a similar material. The area of the panel used for perforations determines the performance of the absorber.

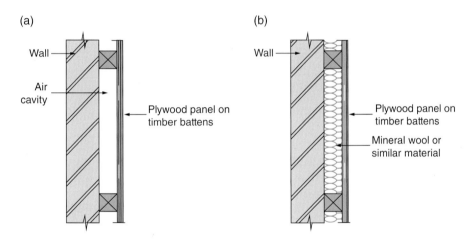

(a)

Wall

Air cavity

Plywood panel on timber battens

(b)

Wall

Plywood panel on timber battens

Mineral wool or similar material

Figure 13.11

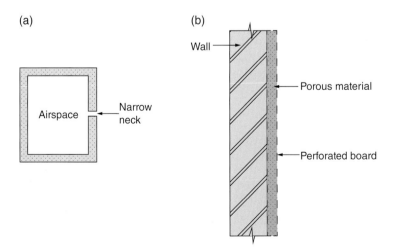

(a)

Airspace

Narrow neck

(b)

Wall

Porous material

Perforated board

Figure 13.12

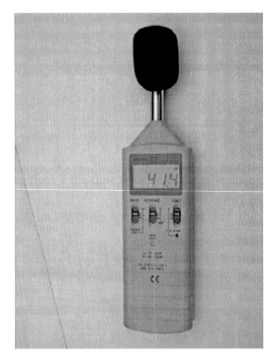

Figure 13.13 Sound-level meter.

Porous materials like mineral wool, glass fibre, curtains and felt, either on their own or provided against a solid backing, will absorb a high-frequency sound of 500 Hz and over.

13.7   Sound-level Meter

The digital sound-level meter, illustrated in Figure 13.13, is a device to measure industrial noise, environmental noise and other sound levels. The sound is detected and sampled by the microphone, and then processed by electronic circuits in the meter. The sound level is finally shown, in decibels, as a digital LCD display. The 'A' frequency weighting can be used for all types of noise, whereas the 'C' frequency weighting is used for measuring the peak value of a noise.

Exercise 13.1

1   A vibrating tuning fork produces a sound wave of 256 Hz frequency and 1.35 m wavelength. Calculate the velocity of sound in air.

2   A musical instrument vibrates at a frequency of 512 Hz. The time for the transmission of vibrations (in air) through a distance of 50 m is 0.15 seconds. Calculate the velocity and wavelength of the sound vibrations in air.

3   Find the velocity of sound in a metal if its density and the modulus of elasticity are 2720 kg/m^3 and 70 kN/mm^2, respectively.

4 The intensity of a sound is 4.5×10^{-7} W/m^2. If the threshold of hearing is 1×10^{-12} W/m^2, find the sound intensity level of this sound in decibels.

5 The noise level produced by a machine is 90 Pa. Find the sound pressure level in decibels if the threshold of hearing (p_0) is 2×10^{-5} Pa.

6 Two machines working at the same time produce noise levels of 6×10^{-1} and 1.3 Pa, respectively. Calculate the overall noise level in decibels.

7 Two power drills and a chainsaw, working at the same time, produce noise levels of 90 dB, 92 dB and 105 dB, respectively. Find the overall noise level using the approximate, and the analytical, method.

8 The intensity of traffic noise incident on a building wall is 5×10^{-5} W/m^2. The wall transmits 8×10^{-8} W/m^2 into the building. Find the sound reduction index of the wall.

9 A hall measuring 40 m × 30 m × 6 m high has the following surface finishes: floor: hardwood blocks on solid floor
   walls: plaster on blockwork
   ceiling: gypsum plaster on concrete
   doors: three – 2 m × 2 m high
   windows: twelve – 1.5 m × 2 m high

   i) Calculate the actual reverberation times at 500 Hz and 2000 Hz if the capacity of the hall is 500 people. Assume that there is a 100% audience and that there is an acoustic shadow on 50% of the floor area.
   ii) Calculate the optimum reverberation time if the hall is to be used for speech.
   iii) Calculate the optimum reverberation time if the hall is to be used for an orchestra.

10 For Question 9, calculate the actual reverberation times at 500 Hz and 2000 Hz if the audience is only 50% of the capacity of the hall. Assume that there is an acoustic shadow on 40% of the floor area.

References/Further Reading

1 DfES (2003). Building Bulletin 93, *Acoustic Design of Schools*.
2 Health and Safety Executive (2005). *Noise at Work – Guidance for Employers on the Control of Noise at Work Regulations 2005*.
3 Tecpel Co. Ltd., Taiwan. Website: www.tecpel.net

14

Light

LEARNING OUTCOMES

1) Define solid angle, luminous intensity, luminous flux and illuminance, and state their units.
2) Explain the inverse square law and the cosine law of illuminance.
3) Perform calculations to determine illuminance directly under, and at a distance away from, a lamp.
4) Use the lumen design method to design the lighting scheme for a room.

14.1 Introduction

Light is a form of energy and forms part of the electromagnetic spectrum. The electromagnetic spectrum ranges from radio waves to gamma rays (Figure 14.1) with light as the only radiation that is visible to the human eye. All **electromagnetic waves** consist of energy in the form of electric and magnetic fields that fluctuate and can transport energy from one location to another. Light waves come in many sizes; the size of a wave is measured as its wavelength. The wavelengths of light waves range from 400 to 700 nanometres (nm).

Many scientists, including Isaac Newton, thought that light was made up of particles of matter that were emitted in all directions from a source. They assumed that light travelled only in straight lines whereas waves bend around obstacles. Robert Hooke and others thought that light was emitted in all directions as a series of waves. Both theories had many flaws but could explain some of the phenomena associated with light. Michael Faraday and James Maxwell proposed that light was a high-frequency electromagnetic radiation that could propagate in the absence of a medium. Finally, in 1905, Albert Einstein's ideas formed the basis for wave-particle dual theory. His theory explained that everything has a particle nature as well as a wave nature.

Isaac Newton discovered that a glass prism, placed in the path of a narrow beam of sunlight, caused a series of overlapping coloured images to be formed on the screen. He called this phenomenon **dispersion of light**, i.e. the splitting of light into its component colours, as shown in Figure 14.2. This shows that sunlight is a mixture of seven colours, the same colours that we find in a rainbow. The coloured band is called a spectrum.

Like all other waves in the electromagnetic spectrum, light can also be reflected, refracted and diffracted. The two types of reflection that are important to interior designers and lighting engineers are explained in Section 14.7.4.

Construction Science and Materials, Second Edition. Surinder Singh Virdi.
© 2017 John Wiley & Sons Ltd. Published 2017 by John Wiley & Sons Ltd.
Companion website: www.wiley.com/go/virdiconstructionscience2e

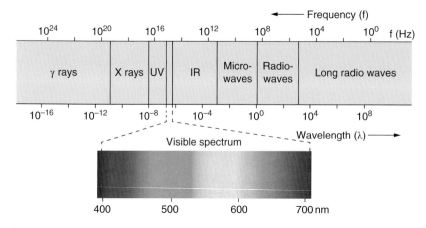

Figure 14.1 Electromagnetic spectrum. (See the colour plate section for a full-colour version of this image.)

Figure 14.2 Dispersion of light.

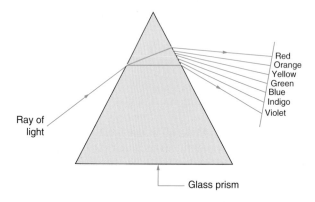

14.2 Additive and Subtractive Colours

White light can be produced by mixing three colours, i.e. red, blue and green, which are known as primary colours. When beams of coloured light are mixed to produce other colours, the process is known as additive colour mixing. For example: Blue + green = cyan (white minus red will also produce the same effect); Green + red = yellow; Red + blue = magenta; Red + blue + green = white (see Figure 14.3).

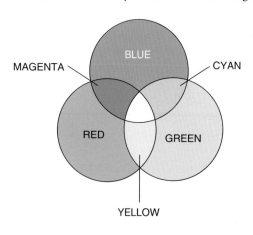

Figure 14.3 (See the colour plate section for a full-colour version of this image.)

Figure 14.4 The radian

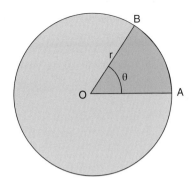

Subtractive colour mixing is based on the concept of taking colours away from white light. For example, a blue surface appears blue when white light falls on it because all the light is absorbed except the blue component, which is reflected. Similarly, a red surface appears red when white light falls on it because all the components are absorbed except the red.

14.3   Measuring Light

14.3.1   Angular Measure

The radian is a unit of measuring two-dimensional angles, for example, angles drawn on paper. In Figure 14.4, arc AB subtends angle θ at the centre of a circle of radius r. The value of angle θ, expressed in radians = $\dfrac{\text{length of arc AB}}{\text{radius}}$.

Angle θ = 1 radian when the length of arc AB equals the radius of the circle.

$$\text{Circumference of the circle} = 2\pi r$$
$$\text{Number of radians in circle} = \frac{\text{Circumference of the circle}}{\text{Radius}}$$
$$= \frac{2\pi r}{r} = 2\pi$$

14.3.2   Solid Angle

A source of light emits the rays of light in all directions. For the measurement of light, it becomes necessary to use a three-dimensional angle through which light rays travel. This three-dimensional angle is known as the **solid angle** (unit: steradian).

Consider a small portion of the surface of the sphere shown in Figure 14.5. By joining the edges of the surface to the centre of the sphere by drawing several lines, a three-dimensional angle is created.

$$\text{The magnitude of the solid angle thus created} = \frac{\text{Area of the surface}}{(\text{Radius})^2} \text{ steradians}$$

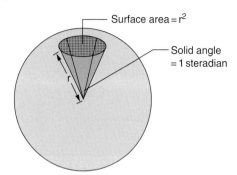

Surface area = r²

Solid angle
= 1 steradian

Figure 14.5 The steradian.

When the surface area is equal to the square of the radius:

Solid angle = 1 steradian

One **steradian** may be defined as the solid angle subtended at the centre of a sphere of radius r by an area of surface equal to the square of the radius.

$$\text{Total surface area of the sphere} = 4\pi r^2$$

$$\text{Total number of steradians in a complete sphere} = \frac{\text{Area of the surface}}{(\text{Radius})^2}$$

$$= \frac{4\pi r^2}{r^2} = 4\pi$$

Therefore, the total solid angle at the centre of a sphere is 4π steradians.

14.3.3 Luminous Intensity (I)

The brightness or intensity of a source of light used to be compared with that of a standard candle. The unit used was known as candle power. The unit of luminous intensity used now is known as candela (cd) and is defined as:

The **luminous intensity** of a source that emits monochromatic radiation of frequency 540×10^{12} Hz and of which the radiant intensity in that direction is $\frac{1}{683}$ watts per steradian.

14.3.4 Luminous Flux (F)

Luminous flux is a measure of the amount of energy emitted in a given time and is defined as the rate at which light energy flows from a source. (Unit: lumen (lm).)

Figure 14.6 shows a point source of light of one candela luminous intensity. The luminous flux emitted within a solid angle of 1 steradian by a point source of light with a uniform luminous intensity of 1 candela is known as 1 lumen, or, in other words, 1 lumen is the quantity of light energy passing per second within the solid angle of 1 steradian.

The total energy radiated from a source is measured in watts. Because we can see only part of this radiation as visible light, we use lumen as the unit of light energy.

The total luminous flux (F) radiated by a uniform source of light with luminous intensity of I candelas, is **4πI** lumens. Mathematically:

$$F = 4\pi I$$

or, $F = SI$, where S is the solid angle.

Figure 14.6 The lumen.

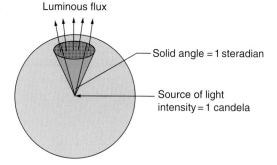

Luminous flux

Solid angle = 1 steradian

Source of light
intensity = 1 candela

Figure 14.7 Summary of definitions.

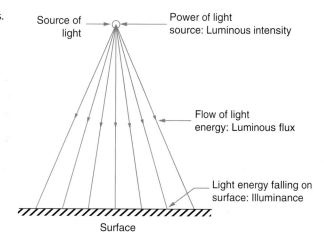

Source of light

Power of light source: Luminous intensity

Flow of light energy: Luminous flux

Light energy falling on surface: Illuminance

Surface

14.3.5 Illuminance (E)

The light energy travels from a source and falls on surfaces to brighten them up. The quantity of light energy falling per second on a unit area of a surface is known as **illuminance**.

$$E = \frac{F}{A}$$

where F is the luminous flux in lumens (lm); A is the surface area in m^2.

The unit of illuminance is lux (lx):

$$1\,\text{lux} = 1\,\text{lm/m}^2$$

Figure 14.7 shows a point source of light with luminous intensity of 1 candela. Assume that the solid angle through which light is emitted is 1 steradian. Then, by definition, the luminous flux emitted by the source is 1 lumen. The luminous flux falls on a surface with an area of 1 m^2 and produces illuminance of 1 lux.

14.3.6 Luminance

The intensity of light emitted by a unit area of a reflecting or luminous surface in a given direction is known as **luminance**. (Unit: cd/m^2.)

Example 14.1   A point source of light has a luminous intensity of 150 cd and radiates in all directions. Calculate the total luminous flux emitted by the source.

Solution:

Total luminous flux, $F = 4\pi I$

$$= 4\pi \times 150 \left(\text{luminous intensity}, I = 150\,\text{cd} \right)$$

$$= \textbf{1884.96 lm}$$

14.4   Inverse Square Law of Illuminance

The illuminance or brightness produced on a surface depends on the distance from the light source. The inverse square law of illuminance states that the illuminance produced on a surface varies inversely as the square of the distance between the source and the surface. The surface is normal to the direction of luminous flux, as shown in Figure 14.8(a).

$$E = \frac{I}{h^2}$$

where E = the illuminance on a surface (lx)
   I = the intensity of the light source (cd)
   h = the perpendicular distance between the light source and the surface.

Figure 14.8(b) shows how illuminance varies as the reciprocal of the square of the distance between the source and the surface. When the surface is 1 m away from the light source (assume I = 720 cd),

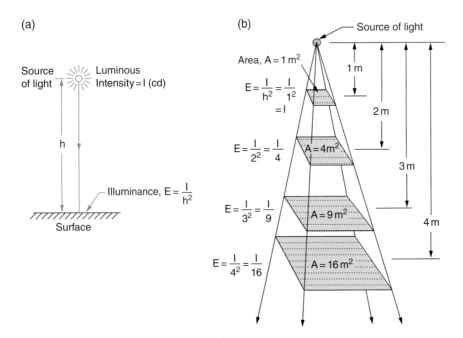

Figure 14.8 Inverse square law.

the illuminance produced is 720 lx. As the distance from the source increases, the flux reaches a bigger area, and hence the illuminance is reduced; for example, at a distance of 2 m:

$$E = \frac{I}{h^2} = \frac{720}{2^2} = 180 \, lx$$

14.5 Lambert's Cosine Law of Illuminance

The inverse square law is only applicable to the points on a surface that receive light rays from a source at right angles. However, lamps emit light at various angles; therefore, the inverse square law needs to be corrected in order to study rays of light that are inclined.

Figure 14.9(a) shows that light falls normally at X, but obliquely (inclined) at Y. The illuminance at X is greater than that at Y because distance h is less than distance d. Also, light falls on a greater area at Y, and therefore results in decreased illuminance.

Consider a beam of light, inclined at angle θ to the normal. Draw line AB normal to the light rays.

$$\text{In } \Delta ABC \left(\text{Figure } 14.9(b) \right), AC = \frac{AB}{\cos \theta} \left(\text{this can also be written as Area } AC = \frac{\text{Area } AB}{\cos \theta} \right)$$

When $\theta = 0°$, $\cos \theta = 1$; therefore $AC = AB = A_1B_1$
When $\theta = 90°$, no point on the floor receives any light, as the light rays are horizontal.
For any angle between 0° and 90°, area AC will be greater than area AB.
When $\theta = 60°$, $\cos \theta = 0.5$; therefore $AC = 2 \, AB$.

The inverse square law $\left(\frac{I}{d^2} \right)$ is modified for light falling obliquely on a horizontal surface and is known as Lambert's cosine law of illuminance:

$$E = \frac{I}{d^2} \cos \theta \tag{14.1}$$

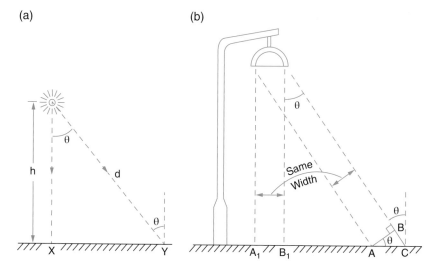

Figure 14.9 Cosine law of illuminance.

From Figure 14.9(a),

$$\frac{h}{d} = \cos\theta, \text{ therefore, } d = \frac{h}{\cos\theta}, \text{ or } d^2 = \frac{h^2}{\cos^2\theta} \tag{14.2}$$

Combining Equations (14.1) and (14.2),

$$E = \frac{I}{h^2}\cos^3\theta \tag{14.3}$$

Equations (14.1) and (14.3) can be used to solve problems that involve light falling obliquely on surfaces.

Example 14.2   A street lamp (point source) with a luminous intensity of 1000 cd is suspended 6 m above the edge of a road. If the road is 5 m wide, calculate the illuminance:

a) At point A on the road surface, which is directly under the lamp.
b) At a point, on the other edge of the road, directly opposite point A.

Solution:
a) I = 1000 cd, h = 6 m

$$\text{Illuminance, } E = \frac{I}{h^2} = \frac{1000}{6^2} = 27.78\,\text{lx}$$

b) From Figure 14.10, $d^2 = 6^2 + 5^2 = 61$

$$d = \sqrt{61} = 7.81\,\text{m}$$

$$\cos\theta = \frac{6}{7.81} = 0.76822$$

$$\text{Illuminance, } E = \frac{I}{h^2}\cos^3\theta$$

$$= \frac{1000}{6^2} \times (0.76822)^3 = 12.59\,\text{lx}$$

Figure 14.10

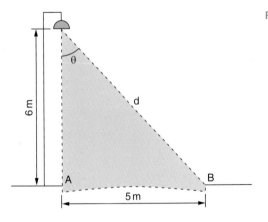

This question can also be solved by using the formula $E = \dfrac{I}{d^2} \cos\theta$

$$E = \frac{1000}{7.81^2} \times 0.76822 = \mathbf{12.59\,lx}$$

Example 14.3 Calculate the illuminance at A due to the two light sources shown in Figure 14.11.

Figure 14.11

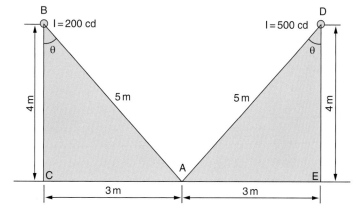

Solution:

$$AB = AD = \sqrt{3^2 + 4^2} = 5\,m$$

$$\cos\theta = \frac{4}{5} = 0.8; h = 4\,m$$

Illuminance at A due to 200 cd source $= \dfrac{I}{h^2} \cos^3\theta$

$$= \frac{200}{4^2} \times 0.8^3 = 6.4\,lx$$

Similarly, illuminance at A due to 500 cd source $= \dfrac{500}{4^2} \times 0.8^3 = 16\,lx$

Total illuminance at $A = 6.4 + 16 = \mathbf{22.4\,lx}$

14.6   Lamps and Luminaires

There are several types of lamp for indoor and outdoor use, but in this section the discussion will be limited to a small selection. The first electric lamp to become popular in domestic and commercial applications is the **incandescent lamp**, which consists of a tungsten filament in a glass enclosure (Figure 14.12(a/b)). When electric current is passed through the filament, it becomes red hot due to its resistance and gives off white light at a temperature of about 2500 °C. The common light bulb, technically known as the general lighting service (**GLS**) lamp, has a coiled tungsten filament.

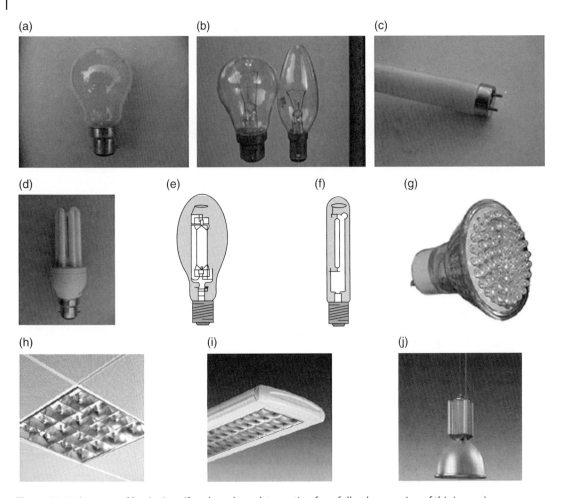

Figure 14.12 Lamps and luminaires. (See the colour plate section for a full-colour version of this image.)

The space in the lamp is filled with inert gases, like argon and nitrogen, to prevent the oxidation of tungsten. The efficacy (ability to convert electrical energy into light energy) is about 10–14 lumens/watt, with an operational life of about 1000 hours. The GLS lamp is cheap, easy to replace and does not require any special device such as a ballast, which is required in fluorescent lamps. But it is the least efficient, as only about 5% of the electrical energy is converted to visible light. In the UK, the production and sale of the GLS lamp are being phased out gradually.

A **fluorescent lamp** or fluorescent tube (MCF) consists of a pair of electrodes, one at each end of a glass tube (Figure 14.12(c)). Mercury vapour and inert gas, like argon, are provided at low pressure inside the glass tube. As a current is passed, the mercury atoms become energised and produce ultra-violet (UV) radiation. The inside of the tube is coated with a chemical compound called phosphor which produces visible light when excited with UV radiation. Fluorescent lamps require a ballast (control gear) to provide a starting pulse of high voltage. A fluorescent lamp is more efficient than an incandescent lamp, having efficacy of 66–94 lm/W (tri-phosphor, linear fluorescent) and an operational life of more than 15000 hours. The **compact fluorescent lamp** (CFL) is a miniature version of the fluorescent tube. Many CFLs are designed to replace GLS lamps and can fit into most existing standard lamp sockets (Figure 14.12(d)). There are two main parts in a CFL: the gas-filled tube and the electronic/magnetic ballast. CFLs consume less electrical energy, as compared to incandescent lamps, to produce the same amount of visible light. They have a longer operational life but are more expensive initially. Their disposal is not easy as they contain mercury. Their luminous efficacy ranges from 50 to 120 lm/W and their operational life ranges from 6000 hours to 15000 hours. Their use in domestic and other applications is becoming more popular despite objections from some people about the possibility of UV radiation and poor light quality.

The **mercury vapour lamp** (Figure 14.12(e)) is a gas discharge lamp that uses mercury in an excited state to produce light. The discharge takes place in a small, fused quartz arc tube provided within a larger glass bulb. Like fluorescent lamps, mercury vapour lamps also utilise a ballast to provide a starting pulse of high voltage to cause ionisation of gas. Mercury vapour lamps are relatively efficient, offer a long operational life and give intense light for use in shops, factories and street lighting.

A **sodium vapour lamp** (Figure 14.12(f)) is also a gas discharge lamp that uses sodium in an excited state to produce light. The low-pressure sodium lamp (SOX) has an inner U-tube containing solid sodium and small amounts of neon and argon gas. The gases start the discharge and produce enough heat to vaporise sodium. These lamps produce yellow light and can be used where colour distinction is not important, as in street lighting and security lighting. High-pressure sodium lamps (SON) are smaller than low-pressure sodium lamps and additionally contain a small quantity of mercury. They give a whiter light and can be used where distinction between colours is important.

LEDs (light emitting diodes) are small, energy-efficient bulbs, illuminated by the movement of electrons in a semiconductor material. They last for several years, give off virtually no heat and do not contain any harmful substances such as mercury. Depending on the material they are made of, they can produce different colours of light. LED bulbs are usually grouped in clusters to produce higher illuminance, as shown in Figure 14.12(g). Their consumption is very low; the following table shows their comparison with incandescent bulbs:

Incandescent bulb	LED	Lumen	Colour
25 W	5 W	>210	White
40 W	7 W	>350	Warm white

A **luminaire** is the light fitting that protects and holds the lamp in position. Made from a range of materials like polycarbonate, aluminium, steel and glass, they may also have reflectors and diffusers, and hold the ballast. Depending on their intended use, they should withstand wet or dry conditions or even corrosive chemicals. Figure 14.12(h)/(i)/(j) show a selection of luminaires.

14.7 Design of Interior Lighting

The design of interior lighting consists of selecting suitable lamps and luminaires to achieve the desired level of illuminance on the working surface (also called the working plane). Many companies have developed sophisticated computer programs to design interior lighting, but the student technician should be able to prepare the design using the properties of lamps and luminaires, room surfaces, the room environment etc. These are discussed in the following sections.

Classification	DLOR	Figure 14.13
Direct	>90%	
Semi-direct	60–90%	
General diffusing	40–60%	
Semi-indirect	10–40%	
Indirect	< 10%	

14.7.1 Light Output Ratio

A lamp is normally provided in a suitable fitting, called the luminaire. The amount of light emitted by a source does not reach the working surface in full; some of the light energy is lost by transmission through the luminaire. The **light output ratio** (LOR) of a luminaire is defined as the percentage of light emitted by it compared with the flux output of the light source.

$$\text{LOR} = \text{Upward LOR}\,(\text{ULOR}) + \text{Downward LOR}\,(\text{DLOR})$$

$$\text{ULOR} = \frac{\text{Total upward light output of the fitting}}{\text{Total light output of the fitting}}$$

$$\text{DLOR} = \frac{\text{Total downward light output of the fitting}}{\text{Total light output of the fitting}}$$

Figure 14.13 shows light fittings with various DLORs.

A source does not emit light equally in all directions, which can be checked by drawing a **polar curve** of the luminaire. The polar curve of a luminaire is drawn by measuring its luminous intensity at different points in a given direction and plotting the points on polar coordinate paper. This fact is considered in the lighting design by taking into account quantities such as the utilisation factor, the downward light output ratio (DLOR) and the direct ratio. The reflectance of room surfaces and the room index also affect the illuminance produced on the working plane.

14.7.2 Direct Ratio

The **direct ratio** is the proportion of the total downward flux from the luminaire that falls directly on the working plane. The plot of direct ratio against room index (explained in Section 14.7.3) gives a curve which is compared to the standard curves and classed as one to which it is closest, i.e. BZ 1, BZ 3 etc. The whole range of possible lamp types is covered by a theoretical set of curves corresponding to ten different light concentrations.

14.7.3   Room Index

The **room index** is also an important factor in calculating the level of illuminance. It is based on the length and the width of the room and on the distance between the lamp and the working plane:

$$\text{Room index}, k = \frac{L \times W}{H_m (L + W)}$$

where L = the length of the room
W = the width of the room
H_m = the height of the light fitting above the working surface.

14.7.4   Reflection of Light

Like all other waves in the electromagnetic spectrum, light can also be reflected. There are two types of reflection that are important to lighting engineers: direct reflection and diffuse reflection.

Direct or **regular reflection** occurs when rays of light are reflected from a smooth surface without scatter. The angle of incidence is equal to the angle of reflection, as shown in Figure 14.14(a).

Diffuse reflection occurs from rough surfaces, as shown in Figure 14.14(b). The rays of light are not parallel after reflection, but for any ray, the angle of incidence is equal to the angle of reflection.

Reflectance is defined as the ratio of the light reflected from a surface to the total light incident on the surface, and is usually expressed as a percentage. Typical values of reflectance for some materials are listed in Table 14.1.

14.7.5   Level of Illuminance

Illumination levels for a wide variety of environments and tasks can be found in The Society of Light and Lighting's *Code for Lighting* and other publications. The levels stated are maintained illuminances, which are the minimum average illumination levels that should be achieved at the point of scheduled maintenance. Table 14.2 shows a selection taken from *The SLL Lighting Handbook.* Some of these values can also be used for other (similar) applications. For example, the level of illuminance in a school canteen is 200 lux; this value can also be used for a restaurant or an office canteen.

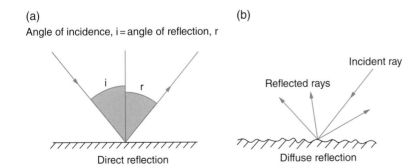

Figure 14.14 Reflection of light.

Table 14.1 Reflectance of building materials.

Surface	Reflectance (%)
Brickwork	10–30
Concrete (new)	40–50
Wallpaper (light-coloured)	55–65
White oil-based paint	75–85
White emulsion (matt)	60–70
Timber (beech, pine)	40–60

Table 14.2 Maintained illuminance. Reproduced by permission of CIBSE.

Building/room	Recommended maintained illuminance (lux)
Educational institutions	
Assembly hall	300
Canteen	200
Classroom (schools)	300
Classroom (colleges)	500
Lecture hall	300
Library	300
Seminar room	300
Science laboratory	500
Industrial buildings	
Corridor, lift, stairs	100
Toilet	100
Workshop	300
Offices	
Conference room	300
Corridor, stairs	100 to 150
Entrance	200
Executive office	300 to 500
Open-plan office	500
Quasi-domestic buildings	
Bedroom, kitchen, lounge	150
Corridor, stairs	100

(Source: CIBSE – *The SLL Lighting Handbook*: 2009)

14.7.6 Utilisation Factor (UF)

The **utilisation factor** is defined as the total flux reaching the working plane compared to the total flux emitted by the luminaire. Utilisation factors can be calculated for any surface or layout of luminaires, but are generally calculated for luminaires arranged in regular arrays. Most manufacturers publish tables of utilisation factors for their luminaires. It is necessary to determine the room index and the reflectance of the room surfaces before finding the utilisation factors from manufacturers' tables. Tables 14.3 to 14.7 show utilisation factors for a selection of luminaires manufactured by Thorn Lighting Company. Similar information is also published by other manufacturers.

Tables 14.3 to 14.5 are based on recessed luminaires with fluorescent tubes, an example of which is shown in Figure 14.12(h). Figure 14.12(j) shows a surface-mounted/suspended fluorescent luminaire. T16 and T26 are the diameters of the fluorescent tube lamps. Table 14.7 is based on recessed luminaires with LED lamps.

14.7.7 Maintenance Factor (MF)

New lamps and luminaires produce maximal light output. After some time, their surfaces will become dirty due to the deposition of dust. A dirty lamp will be unable to emit as much light as when it was clean. The maintenance factor takes into account the loss in illuminance due to various factors that include dirt collecting on lamps, luminaires and room surfaces, and lamp failures:

$$MF = LLMF \times LSF \times LMF \times RSMF$$

LLMF is the **lamp lumen maintenance factor**, and is the proportion of the initial light output that is produced after a specified time. (See Table 14.8.)

Table 14.3 Utilisation factors for 664 × 599 mm recessed modular luminaires (Quattro C Body 4 × 18 W T26 lamps. Flux = 1350 lumens/lamp).

	Utilisation factors								
Room reflectance (%)	Room Index								
Ceiling/Walls/Floor	0.75	1.00	1.25	1.50	2.00	2.50	3.00	4.00	5.00
70/50/20	0.54	0.59	0.62	0.65	0.68	0.70	0.72	0.74	0.75
70/30/20	0.50	0.55	0.59	0.62	0.66	0.68	0.70	0.72	0.73
70/10/20	0.47	0.52	0.57	0.60	0.64	0.66	0.68	0.70	0.72
50/50/20	0.52	0.57	0.61	0.63	0.66	0.68	0.70	0.71	0.72
50/30/20	0.49	0.54	0.58	0.61	0.64	0.66	0.68	0.70	0.71
50/10/20	0.47	0.52	0.56	0.59	0.62	0.65	0.66	0.69	0.70
30/50/20	0.52	0.56	0.60	0.62	0.65	0.66	0.67	0.69	0.70
30/30/20	0.49	0.54	0.57	0.60	0.63	0.65	0.66	0.68	0.69
30/10/20	0.47	0.52	0.55	0.58	0.61	0.63	0.65	0.67	0.68
0/0/0	0.45	0.50	0.54	0.56	0.59	0.61	0.62	0.64	0.65
	DLOR = 69%		ULOR = 0%		SHR Nom = 1.5			SHR Max = 1.66	

(Source: Thorn Lighting Company)

Table 14.4 Utilisation factors for 610×600 mm recessed modular luminaires (Quattro T Body 2×40 W TC-L lamps. Flux = 3500 lumens/lamp).

Room reflectance (%)	Utilisation factors								
	Room Index								
Ceiling/Walls/Floor	0.75	1.00	1.25	1.50	2.00	2.50	3.00	4.00	5.00
70/50/20	0.54	0.59	0.63	0.66	0.69	0.71	0.73	0.75	0.76
70/30/20	0.50	056	0.60	0.63	0.67	0.69	0.71	0.73	0.74
70/10/20	0.47	0.53	0.57	0.60	0.64	0.67	0.69	0.71	0.73
50/50/20	0.53	0.58	0.62	0.64	0.67	0.69	0.70	0.72	0.73
50/30/20	0.50	0.55	0.59	0.61	0.65	0.67	0.69	0.71	0.72
50/10/20	0.47	0.52	0.57	0.59	0.63	0.65	0.67	0.69	0.71
30/50/20	0.52	0.57	0.60	0.63	0.65	0.67	0.68	0.70	0.71
30/30/20	0.49	0.54	0.58	0.60	0.64	0.65	0.67	0.69	0.70
30/10/20	0.47	0.52	0.56	0.59	0.62	0.64	0.66	0.68	0.69
0/0/0	0.46	0.51	0.54	0.57	0.60	0.62	0.63	0.65	0.66

DLOR = 70%   ULOR = 0%   SHR Nom = 1.5   SHR Max = 1.52

(Source: Thorn Lighting Company)

Table 14.5 Utilisation factors for 610×600 mm recessed modular luminaires (Quattro T Body 2×34 W TC-L lamps. Flux = 2800 lumens/lamp).

Room reflectance (%)	Utilisation factors								
	Room Index								
Ceiling/Walls/Floor	0.75	1.00	1.25	1.50	2.00	2.50	3.00	4.00	5.00
70/50/20	0.51	0.56	0.59	0.62	0.65	0.67	0.68	0.70	0.71
70/30/20	0.47	0.52	0.56	0.59	0.63	0.65	0.66	0.69	0.70
70/10/20	0.44	0.50	0.54	0.57	0.60	0.63	0.65	0.67	0.69
50/50/20	0.50	0.54	0.58	0.60	0.63	0.65	0.66	0.68	0.69
50/30/20	0.47	0.52	0.55	0.58	0.61	0.63	0.65	0.66	0.68
50/10/20	0.44	0.49	0.53	0.56	0.59	0.62	0.63	0.65	0.67
30/50/20	0.49	0.53	0.57	0.59	0.61	0.63	0.64	0.65	0.66
30/30/20	0.46	0.51	0.54	0.57	0.60	0.62	0.63	0.64	0.65
30/10/20	0.44	0.49	0.53	0.55	0.58	0.60	0.62	0.63	0.65
0/0/0	0.43	0.48	0.51	0.53	0.56	0.58	0.59	0.61	0.62

DLOR = 65%   ULOR = 0%   SHR Nom = 1.5   SHR Max = 1.52

(Source: Thorn Lighting Company)

Table 14.6 Utilisation factors for 1515 × 217 × 90 mm surface-mounted and suspended fluorescent luminaires (College 2 × 49 WT16 lamps. Flux = 4350 lumens/lamp).

Room reflectance (%)	Utilisation factors								
	Room Index								
Ceiling/Walls/Floor	0.75	1.00	1.25	1.50	2.00	2.50	3.00	4.00	5.00
70/50/20	0.40	0.45	0.50	0.53	0.57	0.61	0.63	0.66	0.68
70/30/20	0.34	0.40	0.45	0.48	0.53	0.57	0.59	0.63	0.65
70/10/20	0.31	0.36	0.41	0.45	0.50	0.54	0.56	0.60	0.63
50/50/20	0.38	0.43	0.47	0.50	0.55	0.57	0.59	0.62	0.64
50/30/20	0.33	0.39	0.43	0.46	0.51	0.54	0.56	0.60	0.62
50/10/20	0.30	0.35	0.40	0.43	0.48	0.52	0.54	0.58	0.60
30/50/20	0.36	0.41	0.45	0.48	0.52	0.54	0.56	0.59	0.60
30/30/20	0.33	0.37	0.42	0.45	0.49	0.52	0.54	0.57	0.59
30/10/20	0.30	0.34	0.39	0.42	0.46	0.50	0.52	0.55	0.57
0/0/0	0.28	0.32	0.36	0.39	0.43	0.46	0.48	0.51	0.53

DLOR = 62%   ULOR = 6%   SHR Nom = 1.5   SHR Max = 1.69

(Source: Thorn Lighting Company)

Table 14.7 Utilisation factors for 597 x 597 x 63 recessed modular luminaires (Elevation LED 3700 HFLI MPT 37 W lamps. Flux = 3700 lumens/lamp). Reproduced by permission of Thorn Lighting Company.

Room reflectance (%)	Utilisation factors								
	Room Index								
Ceiling/Walls/Floor	0.75	1.00	1.25	1.50	2.00	2.50	3.00	4.00	5.00
70 / 50 / 20	0.60	0.70	0.77	0.81	0.88	0.93	0.96	1.00	1.03
70 / 30 / 20	0.53	0.63	0.70	0.75	0.82	0.88	0.91	0.96	0.99
70 / 10 / 20	0.47	0.57	0.64	0.70	0.78	0.83	0.87	0.93	0.96
50 / 50 / 20	0.58	0.67	0.74	0.78	0.84	0.89	0.92	0.96	0.98
50 / 30 / 20	0.52	0.61	0.68	0.73	0.80	0.84	0.88	0.92	0.95
50 / 10 / 20	0.47	0.56	0.63	0.68	0.76	0.81	0.84	0.89	0.99
30 / 50 / 20	0.56	0.65	0.71	0.75	0.81	0.85	0.88	0.91	0.93
30 / 30 / 20	0.51	0.60	0.66	0.71	0.77	0.81	0.84	0.89	0.91
30 / 10 / 20	0.46	0.55	0.62	0.67	0.74	0.78	0.82	0.86	0.89
0 / 0 / 0	0.44	0.53	0.59	0.63	0.70	0.74	0.77	0.81	0.84

DLOR = 96%   ULOR = 4%   SHR Nom = 1.25   SHR Max = 1.43

(Source: Thorn Lighting Company)

Table 14.8 Typical values of LLMF. Reproduced by permission of CIBSE.

Lamp	Operation time								
	2000	4000	6000	8000	10000	12000	14000	30000	Hours
1. Fluorescent tri-phosphor	0.94	0.91	0.87	0.86	0.85	0.84	0.83		
2. Mercury	0.93	0.87	0.80	0.76	0.72	0.68	0.64		
3. High pressure sodium	0.96	0.93	0.91	0.89	0.88	0.87	0.86		
4. LED								0.88	

(Source: The Society of Light and Lighting/CIBSE (for 1, 2, 3))

Table 14.9 Typical values of LMF. Reproduced by permission of CIBSE.

Time between cleaning	1.0			2.0			3.0 Years		
Environment (see Table 14.11)	C	N	D	C	N	D	C	N	D
Luminaire category (see Table 14.11)									
A	0.93	0.89	0.83	0.89	0.84	0.78	0.85	0.79	0.80
B	0.90	0.86	0.83	0.84	0.80	0.75	0.79	0.74	0.79
C	0.89	0.81	0.72	0.80	0.69	0.59	0.74	0.61	0.64
D	0.88	0.82	0.77	0.83	0.77	0.71	0.79	0.73	0.73
E	0.94	0.90	0.86	0.91	0.86	0.81	0.90	0.84	0.83
F	0.86	0.81	0.74	0.77	0.66	0.57	0.70	0.55	0.65

(Source: The Society of Light and Lighting/CIBSE)

LSF is the **lamp survival factor** and takes into account the failure of lamps. It is considered if the lamps are replaced at regular intervals. The value of LSF is 1.0 if the lamps are replaced immediately after their failure.

Dirt deposited on/in the luminaire will cause a reduction in its light output. The amount of dirt deposited at a particular time depends on how clean, or how dirty, the room environment is. The **LMF (luminaire maintenance factor)** is used in the maintenance factor equation to consider the effect of dirt on the light output of luminaires (Table 14.9).

Dirt can also be deposited on room surfaces and can affect the amount of light reflected. This is taken into account by using the **room surface maintenance factor (RSMF)** (Table 14.10).

Table 14.11 shows information on luminaire categories and suitable environments.

14.7.8 Lumen Design Method

Several factors need to be considered before designing a lighting scheme; some of these are:

- Appropriate lamps and luminaires;
- The task that will be performed in the room/building;
- The level of illuminance on the working plane;

Table 14.10 Typical values of RSMF. Reproduced by permission of CIBSE.

Time between cleaning		1.0			2.0			3.0 Years		
Room index K	Luminaire distribution	C	N	D	C	N	D	C	N	D
Small (0.7)	Direct	0.97	0.94	0.93	0.95	0.93	0.90	0.94	0.92	0.98
	Direct/Indirect	0.90	0.86	0.82	0.87	0.82	0.78	0.84	0.79	0.74
	Indirect	0.85	0.78	0.73	0.81	0.73	0.66	0.75	0.68	0.59
Medium to Large (2.5–5.0)	Direct	0.98	0.96	0.95	0.96	0.95	0.94	0.96	0.95	0.94
	Direct/Indirect	0.92	0.88	0.85	0.89	0.85	0.81	0.86	0.82	0.78
	Indirect	0.88	0.82	0.77	0.84	0.77	0.70	0.78	0.72	0.64

(Source: The Society of Light and Lighting/CIBSE)

Table 14.11 Luminaire categories and environmental conditions. Reproduced by permission of CIBSE.

Category	Description	Environment	Typical locations
A	Bare lamp batten	Clean (C)	Clean rooms, computer centres, hospitals
B	Open-top reflector		
C	Closed-top reflector	Normal (N)	Offices, shops, schools, restaurants,
D	Enclosed		warehouses, laboratories
E	Dustproof	Dirty (D)	Steelworks, chemical works, woodwork
F	Indirect uplighter		areas, welding, polishing, foundries

(Source: The Society of Light and Lighting/CIBSE)

- Energy efficiency;
- The problem of glare.

The lumen design method can be used to design lighting schemes in square or rectangular rooms with a regular array of luminaires. The method takes into account the area of the room (**A**), the utilisation factor (**UF**) and the maintenance factor (**MF**) of the lamps/luminaires, luminous flux (**φ**) of lamps and the average illuminance (**E**) required on the working plane. The procedure involves the selection of appropriate lamps/luminaires and then the calculation of the number of luminaires required to produce the necessary level of illuminance.

$$\text{The number of lamps required}, N = \frac{E \times A}{\Phi \times UF \times MF}$$

14.7.9 SHR

The SHR (spacing to height ratio) is the ratio of the distance between the luminaires and their height above the working plane. Most manufacturers publish the values of SHR for their luminaires. From SHR_{max} we can calculate the maximum spacing between the luminaires that can be allowed and still achieve uniformity in illuminance.

Example 14.4 A room measures $7.0\,m \times 6.0\,m \times 2.6\,m$ high. The lamps are surface mounted on the ceiling and the desks are $0.75\,m$ high. Find the room index.

Solution:
The room index is given by the formula:

$$k = \frac{L \times W}{H_m (L + W)} \quad \text{(see also Section 14.7.3)}$$

H_m = height of the light fitting above the working plane
$$= 2.6 - 0.75 = 1.85\,m$$

$$\text{Room index} = \frac{7 \times 6}{1.85 \times (7 + 6)}$$
$$= \frac{42}{1.85 \times (13)} = \mathbf{1.75}$$

Example 14.5 A room measuring $12.0\,m \times 10.0\,m \times 4.0\,m$ high is provided with Quattro T $610 \times 600\,mm$ recessed modular luminaires with $2 \times 40\,W$ TC-L lamps. The height of the working surface is $0.72\,m$. Determine the utilisation and maintenance factors if:

1) The ceiling, wall and floor reflectance are 70%, 50% and 20%, respectively.
2) The lamps are changed when they fail; assume operation time = 10000 hours.
3) The lamps, luminaires and room surfaces are cleaned annually.
4) The room environment is clean.

Solution:

$$\text{Room index} = \frac{12.0 \times 10.0}{3.28 \times (12 + 10)} = 1.66 \quad \left(H_m = 4.0 - 0.72 = 3.28 \right)$$

UF: Refer to Table 14.4. For the given room reflectance and room index of 1.66, the **UF = 0.67**
MF = LLMF $\times$ LSF $\times$ LMF $\times$ RSMF
LLMF = 0.85 (Table 14.8 – for fluorescent tri-phosphor lamps whose operation time is 10000 hours)
LSF = 1 (lamps are changed immediately after their failure)
LMF = 0.89 (Table 14.9 – for clean environment; luminaire category: C; one year time between cleaning)
RSMF = 0.97 (Table 14.10 – for clean environment, luminaire distribution: direct; one year time between cleaning)
Multiplying all the above factors, MF = $0.85 \times 1 \times 0.89 \times 0.97 = \mathbf{0.73}$

Example 14.6 A room measuring $12.2\,m \times 9.7\,m \times 4.2\,m$ high is to be used as a seminar room. The required illuminance on the desks is 500 lux. The $610 \times 600\,mm$ recessed Quattro T Body modular luminaires have been specified for use in the room. Use the lumen design method to determine the number of luminaires required.
   Given:

1) The ceiling, wall and floor surface reflectances are 70%, 50% and 20%, respectively.
2) There is a spot replacement programme for replacing lamps; the operation time of each lamp is 10000 hours.
3) The room environment is clean; the luminaires and the room surfaces are cleaned annually.

Solution:

$$\text{Number of lamps}, N = \frac{E \times A}{\Phi \times UF \times MF}$$

$$\text{Room area}, A = 12.2 \times 9.7 = 118.34 \text{ m}^2$$

$$E = 500 \text{ lx}$$

Use two TC-L/40 W lamps per luminaire. Luminous flux of each lamp = 3500 lm

Luminous flux of each luminaire, $\Phi = 2 \times 3500 = 7000$ lm

$H_m = 4.2 - 0.7 = 3.5$ m(assume the desks are 0.7 m high)

$$\text{Room index} = \frac{L \times W}{H_m (L + W)} = \frac{12.2 \times 9.7}{3.5 \times (12.2 + 9.7)} = 1.54$$

From Table 14.4, UF = 0.66

$$\begin{aligned} MF &= LLMF \times LSF \times LMF \times RSMF \\ &= 0.85 \times 1 \times 0.89 \times 0.97 = 0.73 \end{aligned}$$

(Refer to Tables 14.8, 14.9, 14.10 and Example 14.5)

$$N = \frac{500 \times 118.34}{7000 \times 0.66 \times 0.73} = \mathbf{17.5}$$

The first trial shows that it is not possible to have 18 luminaires in regular arrays. Increase the number of luminaires to 20, as shown in Figure 14.15, to have a regular array. An alternative is to use lamps of different lumen output.

$$SHR = \frac{2.4}{3.5} = 0.69 < SHR_{max}$$

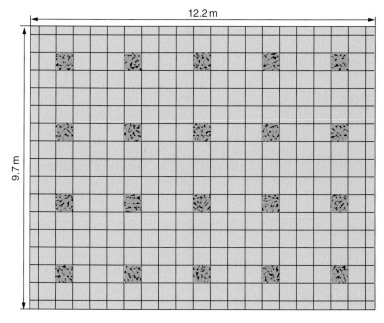

Figure 14.15 Spacing of 600 × 600 mm modular luminaires (shaded squares).

Example 14.7   A room measuring 10.0 m × 8.0 m × 3.5 m high is to be used as an office. The required illuminance on the working plane is 500 lux. Surface-mounted luminaires measuring 1515 × 217 × 90 mm have been specified. Each luminaire will house 2 × 49 W T16 fluorescent lamps, the flux of each lamp being 4350 lumens. Use the lumen design method to determine the number of luminaires required, if:
1) The ceiling, wall and working surface reflectances are 70%, 50% and 20%, respectively.
2) There is a spot replacement programme for replacing lamps; the operation time of each lamp is 10000 hours.
3) The room environment is clean; the luminaires and the room surfaces are cleaned annually.

Solution:

$$\text{Number of lamps, N} = \frac{E \times A}{\Phi \times UF \times MF}$$

$$\text{Room area, A} = 10.0 \times 8.0 = 80.0 \, m^2$$

$$E = 500 \, lx$$

Use two 49 W T16 fluorescent lamps per luminaire. Luminous flux of each lamp = 4350 lm

$$\text{Luminous flux of each luminaire,} \Phi = 2 \times 4350 = 8700 \, lm$$

$$H_m = 3.5 - 0.7 = 2.8 \, m \left(\text{assume the desks are 0.7 m high} \right)$$

$$\text{Room index} = \frac{L \times W}{H_m (L+W)} = \frac{10.0 \times 8.0}{2.8 \times (10.0 + 8.0)} = 1.59$$

From Table 14.6, UF = 0.54

$$MF = LLMF \times LSF \times LMF \times RSMF$$
$$= 0.85 \times 1 \times 0.89 \times 0.97 = 0.73$$

(Refer to Tables 14.8, 14.9, 14.10 and Example 14.5)

$$N = \frac{500 \times 80.0}{8700 \times 0.54 \times 0.73} = 11.7, \text{ say } \mathbf{12}$$

The arrangement of luminaires is shown in Figure 14.16.

Example 14.8   A room measuring 8.7 m × 7.2 m × 3.6 m high is to be used as a seminar room. The required illuminance on 0.72 m high desks is 400 lux. The 597 × 597 × 63 recessed modular LED luminaires (Elevation LED 3700) have been specified for use in the room. Use the lumen design method to determine the number of luminaires required.
Given:
1) The ceiling, wall and floor surface reflectances are 70%, 50% and 20%, respectively.
2) There is a spot replacement programme for replacing lamps. The operation time of each lamp is 30000 hours.
3) The room environment is clean; the luminaires and the room surfaces are cleaned annually.

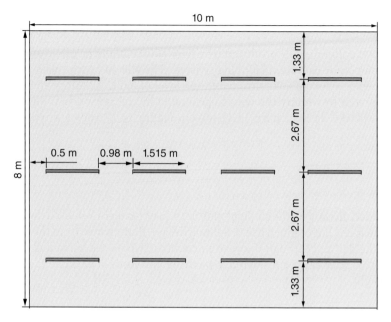

Figure 14.16 Spacing of 1515 × 217 mm modular luminaires.

Solution:

$$\text{Number of lamps, N} = \frac{E \times A}{\Phi \times UF \times MF}$$

$$\text{Room area, A} = 8.7 \times 7.2 = 62.64 \text{ m}^2$$

$$E = 400 \text{ lx}$$

$$\text{Luminous flux of each lamp} = 3700 \text{ lm}$$

$$\text{Luminous flux of each luminaire, } \Phi = 3700 \text{ lm}$$

$$H_m = 3.6 - 0.72 = 2.88 \text{ m}$$

$$\text{Room index} = \frac{L \times W}{H_m (L + W)} = \frac{8.7 \times 7.2}{2.88 \times (8.7 + 7.2)} = 1.37$$

From Table 14.7, UF = 0.79

$$MF = LLMF \times LSF \times LMF \times RSMF$$
$$= 0.88 \times 1 \times 0.89 \times 0.97 = 0.76$$

(Refer to Tables 14.8, 14.9 and 14.10 and Example 14.5)

$$N = \frac{400 \times 62.64}{3700 \times 0.79 \times 0.76} = 11.28$$

Provide 12 luminaires.

$$SHR = \frac{2.4}{2.88} = 0.87 < SHR_{max} \quad (SHR_{max} = 1.43)$$

14.8   Light Meter

A light meter is used to measure the illuminance produced by a light source in lux or other units. Most modern light meters use silicon or cadmium sulphide sensors. The silicon sensor, which is connected to the meter, will generate a voltage proportional to the level of light. A battery and an amplification circuit are incorporated to amplify the voltage generated by the sensor. In a digital light meter, the light level is displayed as lux on an LCD screen. Figure 14.17 shows a typical light meter.

14.9   Daylighting

Daylight, or natural light, is provided directly by the sunlight and/or the sky. Sunlight, which is direct light from the sun, is so variable that it is not of much use in lighting design. Illuminance from the sky is also variable; it varies during the day and through the year. In countries where the climate is not reliable (for example, the UK), an overcast sky is considered to be the main source of daylight. It also represents the worst type of daylight.

14.9.1   Uniform Sky

A uniform sky is an overcast sky considered to have the same luminance for the entire hemispherical sky.

14.9.2   CIE Standard Overcast Sky

In the CIE (Commission Internationale de l'Eclairage) sky, the luminance varies with the angle above the horizon. This has a close resemblance to the luminance distribution of a real overcast sky.
   Luminance at an angle θ (L_θ) from the horizon, as shown in Figure 14.18, is given by:

$$L_\theta = L\left(\frac{1 + 2\sin\theta}{3}\right)$$

where L is the luminance at the zenith.
   The luminance at the zenith is three times the luminance at the horizon.

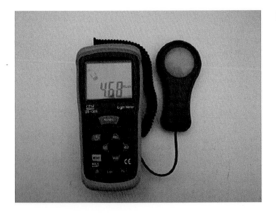

Figure 14.17 Light meter.

Figure 14.18

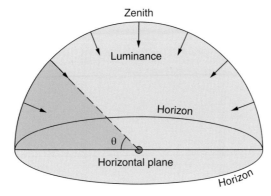

14.9.3 Daylight Factor

Daylight enters the interior of a building through windows or roof lights. The illuminance due to daylight inside a building will reduce rapidly as the distance from the side window/patio door increases. The amount of daylight that is able to penetrate to a point in a building is expressed as the daylight factor.

The **daylight factor** is defined as the ratio of the natural illuminance at a particular point on a horizontal plane to the simultaneously occurring external illuminance of the unobstructed overcast sky. In the UK, the standard sky is assumed to give at least 5000 lx of illuminance on the ground.

$$\text{Daylight factor} \left(\text{DF} \right) = \frac{\text{Internal illuminance}}{\text{External illuminance}} \times 100$$

Components of the Daylight Factor

The daylight reaching any point inside a room is made up of three components (Figure 14.19):

1) **Sky component (SC):** Light received directly from the sky through windows. Direct sunlight is not considered for calculating the sky component.

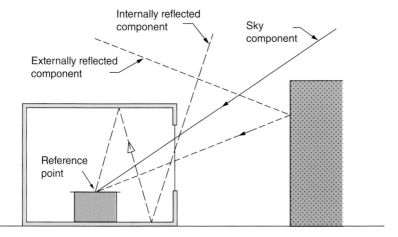

Figure 14.19 Components of the daylight factor.

(a)   (b)

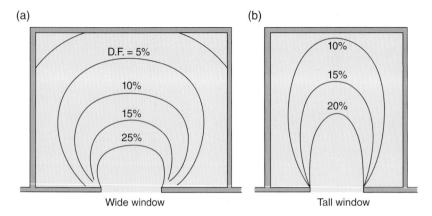

Wide window   Tall window

Figure 14.20 Daylight factor contours.

2) **Externally reflected component (ERC):** Light received directly from external reflecting surfaces like buildings, vegetation etc.
3) **Internally reflected component (IRC):** Light received after multiple reflections off the floor, walls and ceiling.

The daylight factor is the sum of the three components:

$$DF = SC + ERC + IRC$$

The sky component and the internally reflected component are the more important components; the externally reflected component makes a very small contribution to the daylight factor.

In side-lit rooms, the maximum DF is near the windows, and it is mainly due to the sky component. Both the sky component and the externally reflected component decrease as the distance from a window increases. The internally reflected component, however, remains uniform in the room's interior.

The daylight contours, which are lines on a plan joining all points of the same DF, show how the daylight factor reduces away from a window. Compared to wide windows, tall windows provide greater penetration of daylight, as shown in Figure 14.20.

Example 14.9   Calculate the illuminance at a point in a building if a daylight factor of 3% is required. Assume that the external illuminance due to an unobstructed standard sky is 5000 lx.

Solution:

$$DF = \frac{\text{Internal illuminance}}{\text{External illuminance}} \times 100$$

Transpose the formula to make Internal illuminance the subject:

$$\text{Internal illuminance} = \frac{DF \times \text{External illuminance}}{100}$$

$$= \frac{3 \times 5000}{100} = \mathbf{150\,lx}$$

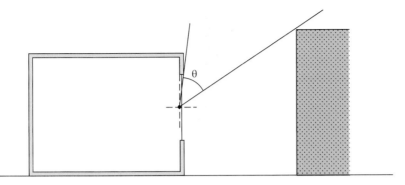

Figure 14.21 Visible sky angle.

Prediction of Daylight Factors

Several methods are available to predict daylight factor components and total daylight factors:

- BRS daylight protractors;
- BRS simplified daylight tables;
- IRC nomograms/IRC formula;
- Pilkington sky dots;
- Waldram diagrams;
- Computer programs.

Publications by the Building Research Establishment and other organisations give detailed descriptions of these methods.

In the early stages of building design, the average daylight factor may be used to assess the adequacy of daylight:

$$\text{Average DF} = \frac{W}{A} \frac{T\theta}{\left(1 - R^2\right)}$$

where: W = the net glazed area of the windows (m^2)
$\quad\quad\quad A$ = the total area of the internal surfaces (m^2)
$\quad\quad\quad T$ = the glass transmittance corrected for dirt
$\quad\quad\quad \theta$ = the visible sky angle in degrees (see Figure 14.21)
$\quad\quad\quad R$ = the average reflectance of area A

The values of W, T and R can be obtained from the BS Daylight Code.

The recommended minimum average daylight factor for dwellings varies between 1 and 2%. This represents an illuminance of between 50 and 100 lx, which is easily exceeded on a bright day.

Exercise 14.1

1   A point source of light has a luminous intensity of 250 cd and radiates in all directions. Calculate the total luminous flux emitted by the source.

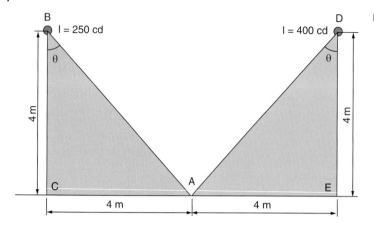

Figure 14.22

2   A street lamp (point source) with a luminous intensity of 1200 cd is suspended 6 m above the edge of a road. If the road is 8 m wide, calculate the illuminance:
  i)   At point A on the road surface, which is directly under the lamp.
  ii) At a point, on the other edge of the road, directly opposite point A.

3   Calculate the illuminance at A due to the two light sources shown in Figure 14.22.

4   A room measures 7.0 m × 6.0 m × 2.8 m high. The lamps are surface mounted on the ceiling and the desks are 0.75 m high. Find the room index.

5   A room measuring 14.0 m × 10.0 m × 4.2 m high is provided with Quattro T 610 × 600 mm recessed modular luminaires with 2 × 40 W TC-L lamps. The height of the working surface is 0.70 m. Determine the utilisation and maintenance factors if:
  i)   The ceiling, wall and floor reflectances are 50%, 50% and 20%, respectively.
  ii) The lamps are changed when they fail; assume operation time = 10000 hours.
  iii) The lamps, luminaires and room surfaces are cleaned every two years.
  iv) The room environment is clean.

6   A room measuring 12.0 m × 10.0 m × 4.0 m high is to be used as a seminar room. The required illuminance on the desks is 500 lux. The 610 × 600 mm recessed Quattro T Body modular luminaires have been specified for use in the room. Use the lumen design method to determine the number of luminaires required, if:
  i)   The ceiling, wall and floor reflectances are 70%, 50% and 20%, respectively.
  ii) There is a spot replacement programme for replacing lamps; the operation time of each lamp is 10000 hours.
  iii) The room environment is clean; the luminaires and the room surfaces are cleaned annually.

7   A room measuring 15.0 m × 12.0 m × 3.5 m high is to be used as an office. The required illuminance on the working plane is 500 lux. Surface-mounted luminaires measuring 1515 × 217 × 90 mm have been specified. Each luminaire will house 2 × 49 W T16 fluorescent lamps, the flux of each lamp being 4350 lumens. Use the lumen design method to determine the number of luminaires required, if:

i) The ceiling, wall and working surface reflectances are 70%, 50% and 20%, respectively.
ii) There is a spot replacement programme for replacing lamps; the operation time of each lamp is 10000 hours.
iii) The room environment is clean; the luminaires and the room surfaces are cleaned annually.

8   Calculate the illuminance at a point in a building if a daylight factor of 2.5% is required. Assume that the external illuminance due to an unobstructed standard sky is 5000 lx.

References/Further Reading

1. CIBSE/The Society of Light and Lighting (2002). *Code for Lighting.*
2. CIBSE/The Society of Light and Lighting (2009). *The SLL Lighting Handbook.*
3. Jones, M., Tingle, M., Petheram, L. and Gadd, K. (2005). *AQA GCSE Science.* London: Harper Collins.
4. Thorn Lighting Company. Website: www.thornlighting.co.uk

15

Human Comfort

15.1 Introduction

Human comfort inside a building is important for its inhabitants to carry out their activities in a safe and efficient manner. A number of factors are important to create a comfortable environment within a building; these are:

- Temperature;
- Air movement;
- Humidity;
- Ventilation;
- Noise;
- Lighting.

15.2 Temperature

A building should be maintained at an appropriate temperature so that the occupants feel comfortable. The food we consume results in the production of heat energy, which is constantly lost through the skin and through bodily functions like breathing. The heat gain and heat loss must be balanced to maintain the inner body temperature at $37 \pm 0.5\,°C$. If the temperature of the surroundings in a building is low, heat loss from the human body will be excessive and could cause health problems such as cold, fever, hypothermia etc. It is, therefore, important to maintain the air temperature in a building within reasonable limits.

Despite large variations in the temperature of the external environment, the human body temperature is maintained within a range of values around a set point of $37\,°C$. The process of maintaining constant conditions in the internal environment of the body is known as **homeostasis**, and when applied to temperature it is called **thermoregulation**.

Construction Science and Materials, Second Edition. Surinder Singh Virdi.
© 2017 John Wiley & Sons Ltd. Published 2017 by John Wiley & Sons Ltd.
Companion website: www.wiley.com/go/virdiconstructionscience2e

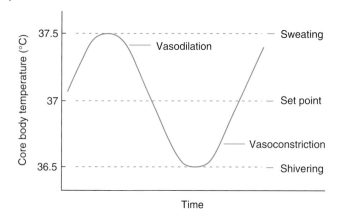

Figure 15.1 Thermoregulation.

There are special sensors in our body that sense the temperature of the blood and send this information to the brain, which knows that the proper temperature should be around $37 \pm 0.5\,°C$. If the sensed temperature is significantly higher, signals are sent to the sweat glands of the skin and surface blood vessels. Sweat glands produce sweat, which, on evaporation, causes cooling. Blood vessels dilate to allow more heat loss through the skin. If the external temperature is low, signals are sent to the muscles to cause shivering (to produce heat), and for the surface blood vessels to constrict (Figure 15.1).

Although not stated by law, the temperature in work rooms should normally be:

- At least $16\,°C$, or
- $13\,°C$ if much of the work involves rigorous physical effort.

In factories where temperatures are usually high, it is possible to work safely provided appropriate controls are present. The interaction between radiant temperature, humidity, air velocity and other factors becomes more complex with rising temperature. The Workplace (Health, Safety and Welfare) Regulations 1992 lay down particular requirements for most aspects of the working environment. Regulation 7 deals specifically with the temperature in indoor workplaces.

The factors that affect thermal comfort in a building are:

- Air temperature;
- Mean radiant temperature;
- Ventilation and air flow;
- Humidity;
- Activity level;
- Clothing.

15.2.1   Air Temperature

The **air temperature**, or dry bulb temperature, is the temperature of air in a room, which is independent of air movement and heat radiation from the surroundings. Air temperature can be measured by mercury thermometer, digital thermometer, thermocouple or other devices. The thermometer should be shielded to protect it from radiation from the surrounding surfaces.

15.2.2   Mean Radiant Temperature

The **mean radiant temperature** (t_r) is the average temperature of the room surfaces. Thermal comfort in a room depends
on the exchange of radiation between the human body and the
surrounding surfaces. The radiation exchange depends on the
size, nature and temperature of the surfaces.

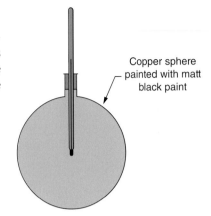

Copper sphere
painted with matt
black paint

$$t_r = \frac{a_1 t_1 + a_2 t_2 + a_3 t_3 + \ldots\ldots\ldots}{a_1 + a_2 + a_3 + \ldots\ldots\ldots}$$

where t_1, t_2 and t_3 are surface temperatures (°C)
a_1, a_2 and a_3 are surface areas (m^2)

A **globe thermometer** may be used to estimate the mean
radiant temperature within a room. It consists of a mercury
thermometer with its bulb at the centre of a 150 mm diameter

Figure 15.2 Globe thermometer.

blackened copper globe, as shown in Figure 15.2. The globe temperature (t_g) lies between the air
temperature and the mean radiant temperature. In still air (velocity, $v = 0$), the globe temperature and
the mean radiant temperature are equal.

15.2.3   Environmental Temperature

Environmental temperatures are used in the calculation of U-values of building elements. Basically,
they are the assumed temperatures, inside and outside a building. The inside temperature, t_{ei}, is a
combination of radiant and air temperatures:

$$t_{ei} = \frac{2}{3}t_r + \frac{1}{3}t_{ai}$$

where t_r = the mean radiant temperature (°C)
t_{ai} = the inside air temperature (°C)

15.2.4   Dry Resultant Temperature

Dry resultant temperature (t_{res}) has been adopted by CIBSE as a thermal index for moderate thermal
environments. Like the environmental temperature, this is also a combination of the inside air temperature and the mean radiant temperature:

$$t_{res} = \frac{t_r + t_{ai}\sqrt{10v}}{1 + \sqrt{10v}}$$

where v = the indoor air speed (m/s)

$$t_{res} = \frac{1}{2}t_r + \frac{1}{2}t_{ai} \text{ when indoor air speed} = 0.1 \text{ m/s}$$

Table 15.1 shows the recommended (CIBSE) dry resultant temperatures for some buildings.

Table 15.1 Typical dry resultant temperatures.

Building/room type	Dry resultant temperature (°C)
Dwellings:	
Bedrooms	17–19
Kitchens	17–19
Living rooms	22–23
Toilets	19–21
Educational buildings:	
Lecture halls	19–21
Seminar rooms	19–21
Factories:	
Heavy work	11–14
Light work	16–19
Offices	21–23
Shops, supermarkets	19–21

(Reproduced from CIBSE Guide A: Environmental Design)

15.2.5 Activity

The metabolic rate of a person depends on their level of activity. Metabolism may be defined as the sum of all chemical reactions that take place in the human body for converting the food consumed into energy. The metabolic rate, and hence the heat generated by a person, increases as the rate of expenditure of physical energy increases. Thus, a person will be comfortable at a slightly lower resultant temperature. Sedentary people lose body heat at a rate of about 120 W at a temperature of 21 °C. The rate of body heat loss will increase to about 240 W on taking up light physical activity. The room temperature may be reduced slightly to maintain the level of thermal comfort.

15.2.6 Clothing

Clothes provide thermal insulation due to the low thermal conductivity of the clothing material and the air layer trapped between the clothing and the skin. The thermal resistance of clothing is expressed in terms of a unit known as the **clo-value**: 1 clo being the insulation provided by such clothing as a business suit. In winter, people wear thicker and heavier clothing to achieve better insulation, and hence more comfort. In summer, people need to lose more heat from their bodies; therefore, lightweight clothing such as a shirt/blouse and trousers/skirt may provide the necessary comfort. Women, generally, wear lighter clothing than men and hence prefer slightly higher temperatures.

15.3 Air Movement

Air movement in a building is usually due to ventilation and draughts. The draughts could be due to gaps around openable parts of doors and single-glazed windows. Air movement creates freshness in a building and is more desirable in summer than in winter. Air movement increases heat loss from the human body and causes chilliness and discomfort in winter. In domestic buildings, an air velocity of 0.1–0.2 m/s is considered to be reasonable. As a draught increases and an air velocity of 0.2 m/s or more is achieved, an increase in the dry resultant temperature is required to compensate for the cooling effect produced by the air movement.

15.4 Humidity

The effect of humidity on human comfort is studied by considering the relative humidity (R.H.) of air (see Chapter 8 for details on R.H.). Its effect is considered to be very small when the resultant temperature is close to the recommended value. At temperatures above 25 °C (approximately) and from higher activity level, heat is lost from the body by sweating. If the R.H. of the air is high, i.e. > 70%, the evaporation of sweat is very slow, so that people feel uncomfortable. The comfortable range of relative humidity is about 40–70%. If the R.H. of air is below 40%, some people may suffer from a dry throat and dryness of the skin and eyes.

15.5 Ventilation

Ventilation is the process by which clean air is brought into a building to remove the stale air. It is essential in buildings to remove carbon dioxide gas, body odours, bacteria, cooking smells and humidity. In crowded rooms, the humidity may increase due to respiration and perspiration of the occupants, and there will also be an increase in the air temperature. Adequate ventilation will bring in fresh air to maintain the supply of oxygen for breathing, remove smells and excess water vapour and lower the air temperature. A selection of ventilation requirements is shown in Table 15.2.

15.6 Predicted Mean Vote

The thermal comfort of people in a building depends on the combined effect of six factors, i.e. air temperature, mean radiant temperature, humidity, air velocity, metabolism rates and clothing level. When a group of people is subject to the same thermal environment, each person experiences the

Table 15.2 Extract ventilation rates.

Building/room type	Minimum intermittent extract rate (litres/s)
Dwellings:	
Bathrooms	15
Kitchens	60
Sanitary accommodation	6
Utility room	30
Offices	Whole building ventilation 10 litres/s per person

(Source: Building Regulations 2000, Approved Document F)

thermal sensation produced by these factors differently. Ideally, the thermal environment in a building should be such that the highest percentage of people is thermally comfortable. The 'predicted mean vote' (PMV) index predicts the average vote of a large group of people on a seven-point thermal sensation scale: −3 for cold, zero for neutral and +3 for hot.

The predicted percentage of dissatisfied (PPD) index predicts the percentage of people that will be dissatisfied with the thermal environment. The PPD increases as the PMV moves away from the neutral position on the thermal sensation scale.

15.7 Noise

Noise can be defined as unwanted sound. Noise in a building may be produced by a multitude of sources like printers, photocopiers, air-conditioning systems, people in a room and people in adjoining rooms. Whatever the source, noise may interfere with our hearing of speech, distract us from what we are trying to do or it may just be annoying.

Various criteria have been developed to analyse noise and hence minimise, or avoid, disturbance and speech interference. Two of these criteria, i.e. speech interference criteria and noise criteria, and their curves are explained here.

In the presence of background noise, verbal communication is affected; Figure 15.3 shows the level at which we must speak to be heard clearly. If the distance between the speaker and the listener is small, then higher levels of background noise may be acceptable. As the distance between them increases, the acceptable levels of background noise must decrease for the speaker to be heard clearly.

Noise criterion (NC) curves were developed in the USA for rating indoor noise, for example, noise from air-conditioning systems. Sound intensity levels that were acceptable to people working in a wide variety of environments were used to produce the NC curves. For a given noise spectrum, the NC rating can be obtained by plotting its octave band levels on the set of NC curves. The NC curve that is just above the noise spectrum determines the noise criterion for the building/space. Figure 15.4(a) shows the set of NC curves.

Noise rating (NR) curves, shown in Figure 15.4(b), are commonly used in Europe for specifying noise levels from mechanical services. NC curves are very similar to the NR curves at middle frequencies.

Table 15.3 shows a selection of the recommended (maximum) values of noise level.

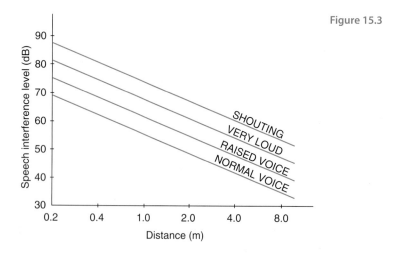

Figure 15.3

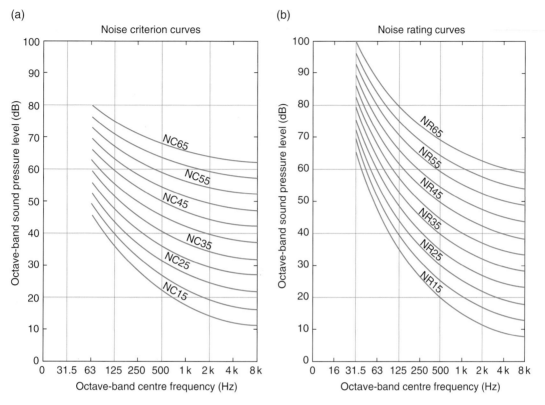

Figure 15.4 (a) Noise criterion curves; (b) noise rating curves. Reproduced by permission of CIBSE.

Table 15.3 Recommended maximum noise levels.

Building/space	Noise rating (NR)
Classrooms	25–35
Lecture rooms	25–35
Libraries	30–35
Dwellings: bedrooms	25
living rooms	30
Hospitals	30–35
General offices	35
Supermarkets	40–50

(Reproduced from CIBSE Guide A: Environmental Design)

15.8 Lighting

Light is a form of energy that forms the visible part of the electromagnetic spectrum (see Chapter 14 for details). Lighting in a building should be adequate so that the tasks may be carried out without any discomfort. In large buildings like office blocks, public libraries and colleges/universities, daylight cannot provide an adequate level of illuminance, therefore artificial lighting is used irrespective of the time of day. In dwelling houses, illuminance during the day is provided mainly by daylight.

One of the problems with daylight is how we can admit enough of it without creating uncomfortable glare and heat. Direct sunlight can cause discomfort through heating and glare, but is often desirable in winter months. Larger windows, although desirable for letting in more daylight, also cause greater heat loss in winter.

The illuminance requirement in a building depends on the type of work to be done: Table 14.2 (Chapter 14) gives the recommended levels of illuminance for some buildings. Poor lighting does not provide adequate illuminance and hence people will find it difficult to do precise work. For example, in an electronic assembly factory, adequate illuminance is very important for precision. Very bright light, on the other hand, could produce eye strain and glare. **Glare** is a sensation produced by luminaires, windows and other objects, seen directly or by reflection, that are too bright compared with the general brightness to which the eyes have adapted. The usual effects of excessive glare in the office include eye strain, visual fatigue or similar visual discomfort. Glare can be of two types: disability glare and discomfort glare. Disability glare occurs when, in the presence of a very bright object (lamp or reflecting surface), less bright objects cannot be seen properly. Discomfort glare causes visual discomfort without reducing the ability to see the details of an object. Discomfort glare is more likely to occur in a building as compared to disability glare.

The problem of glare can be tackled by reducing the intensity of the light source and by moving the source beyond the line of sight.

References/Further Reading

1 CIBSE Guide A (1999). *Environmental Design*. London: Chartered Institution of Building Services Engineers.
2 CIBSE Guide A (2006). *Environmental Design*. London: Chartered Institution of Building Services Engineers.
3 Health and Safety Executive (2007). *Workplace Health, Safety and Welfare – A short guide for managers*.
4 *The Building Regulations* (2000). *Approved Document F – Means of Ventilation*. London: Department of Communities and Local Government.

16

Construction Materials

LEARNING OUTCOMES

1) Describe how bricks, cement, concrete, steel, plastics and glass are manufactured.
2) Describe the main properties of bricks, concrete blocks, cement, concrete, ferrous/non-ferrous metals, timber, plastics and glass.
3) Explain how the above materials can deteriorate and discuss preventative treatments.
4) Discuss the impact on the environment of extracting raw materials and the manufacturing processes of bricks, cement and other materials.

16.1   Introduction

The construction of buildings and civil engineering structures involves the use of a large range of materials, for example, in buildings we use bricks, concrete, cement, aerated concrete blocks, timber, steel, plaster and other materials. In this chapter, the salient features of a small selection of construction materials will be discussed.

Before a material is specified for a particular building element, the architect/designer has to consider the functional requirements of the building element and ensure that the properties of the selected material will fulfil those requirements. The properties of a selection of materials are described in Sections 16.2 to 16.9, but their definitions are given below:

- **Appearance:** Exposed materials must be aesthetically appealing to us. The colour and texture of a material contribute to its appearance.
- **Density:** Density is defined as the mass per unit volume (kg/m^3). Dense materials have high strength but poor thermal insulation whereas lightweight materials have low strength but good thermal insulation.
- **Durability:** Durability is the ability of a material to withstand the damaging effects of force, weathering action, corrosion, chemical attack and living organisms for a given or long period of time while maintaining its desired physical and other properties.
- **Ductility:** The ability of a material to deform plastically without fracture under a tensile force is known as ductility.
- **Malleability:** Malleability is the property of a material by virtue of which it can be deformed by compressive force without cracking or rupturing.
- **Fire resistance:** Fire resistance is the ability of a material to withstand fire and continue to perform structurally for a specified temperature and time.

Construction Science and Materials, Second Edition. Surinder Singh Virdi.
© 2017 John Wiley & Sons Ltd. Published 2017 by John Wiley & Sons Ltd.
Companion website: www.wiley.com/go/virdiconstructionscience2e

- **Hardness:** The resistance of a metal to indentation is known as hardness. The term may also apply to resistance to scratching, cutting or abrasion.
- **Sound insulation/absorption:** Sound insulation is the reduction in the transmission of sound either from one part of a building to another or from external sources to the interior of the building. Sound absorption refers to the absorption of sound energy by a material or object when sound waves strike them. The coefficient of sound absorption is a measure of how good a material is at absorbing sound, and it is defined as the ratio of the amount of sound energy absorbed by a material to the amount of the incident sound energy.
- **Stiffness:** When a force is applied to a material, it tries to produce deformation in that material. Stiffness is the resistance of the material to such deformation.
- **Strength:** The maximal resistance that a material can offer to tensile, compressive, bending and other stresses is known as its strength. Units: N/mm^2, kN/mm^2.
- **Thermal conductivity (λ):** Thermal conductivity is the ability of a material to conduct heat. It is defined as the quantity of heat transmitted through 1 m thick material with a surface area of 1 m^2, due to a temperature difference of 1 °C. The units in the SI system are W/m K or W/m °C.
- **Thermal movement:** A change in the dimensions of a material resulting from fluctuations in temperature over time is known as thermal movement.
  Coefficient of linear expansion can be used to predict the change in the length of a solid material due to an alteration in temperature. It is defined as the change in the length of a material one unit long when its temperature is changed by 1 °C. For more details refer to Chapter 7.
- **Young's modulus:** Also known as the modulus of elasticity, it is defined as the ratio of stress along an axis to strain along that axis. Young's modulus is a measure of the stiffness of a material.

16.2 Bricks

Bricks have been in use for thousands of years for constructing the substructure and the superstructure of buildings. They were first used in the form of sun-dried bricks but later the Romans found that by firing them, their strength could be increased considerably. Although the process has become more mechanised now, the basic method of making bricks has not fundamentally changed. Bricks are now used in building construction internally as well as externally due to their high strength and durability. Bricks, which are made from clay, sand and lime, or from concrete, may be defined as building units that are easy to handle.

16.2.1 Clay Bricks

Clay is a natural material formed due to the weathering of rocks and consists mainly of two minerals: silica (SiO_2) and alumina (Al_2O_3). Smaller amounts of iron oxide, magnesium oxide and chalk may also be present. For brick-making, clay must be plastic, have sufficient strength on drying and its particles must fuse together when subjected to a temperature of over 900 °C. Clay is extracted from pits and quarries and transported to the works.

16.2.2 Size

The size of clay bricks has evolved over many centuries, but still it varies from one country to another. In the United Kingdom, the standard size of clay bricks is $215 \times 102.5 \times 65$ mm.

Clay usually shrinks on drying and firing, therefore the brick manufacturers allow for this shrinkage by making the bricks originally bigger than the finished size after firing. However, it is not possible to produce bricks of identical size because of variations in the properties of various clay soils.

The standard size (i.e. $215 \times 102.5 \times 65$ mm) is the average size of a single brick based on a sample of 24 bricks.

16.2.3 Classification

Clay bricks are classified by British Standards (BS 3921:1985) by considering variety, quality and type:

- **Variety:** There are three main varieties: engineering, facing and common. Engineering bricks are the strongest and are used where the strength of brickwork is very important, for example, the construction of retaining walls, inspection chambers, substructures etc. These are usually red or blue in appearance. Facing bricks, which are available in several colours and textures, are made for appearance and resistance to weather. Common bricks are the least expensive and made for internal work mainly but can be used for external work where rendered.
- **Quality:**
  - Internal quality (to be used internally).
  - Ordinary quality (bricks are durable enough to be used in constructing external walls).
  - Special quality (intended for use in harsh and exposed situations).
- **Type:**
  - Solid: Having no holes or cavities or depressions.
  - Perforated: The volume of holes should be less than 25% of the total brick volume.
  - Frogged: Having depressions in one or more bed faces, the total not exceeding 20% of the total brick volume.

16.2.4 Manufacture

The manufacture of clay bricks starts with the **extraction** of clay (the raw material) from pits and quarries. The raw material is transported to the works where the other processes of manufacture are undertaken. The next stage, called the **preparation** of clay, involves the removal of stones and breaking up of large lumps of clay. Water is added to make the clay–water mixture plastic. The clay–water mixture is then moulded into bricks, this process being known as **forming**. The three main methods of forming bricks are:

- **Soft mud process:** Done by hand as the higher amount of water in the clay makes this possible. The clay is pressed into moulds which are sanded for easy removal of the green brick.
- **Wire cut process:** Used on moderately stiff clay, this involves forcing the clay through a die. The bricks are cut by wire to suitable lengths.
- **Stiff plastic process:** Used on stiff clay, this involves pressing the clay into moulds mechanically.

The green bricks are allowed to dry out before firing in the kiln. The firing temperature for most bricks is 1000 °C; however, engineering bricks are fired at higher temperatures. After firing, the bricks are allowed to cool before being packaged.

Figure 16.1 shows all stages of brick manufacture.

16.2.5 Properties

The properties of clay bricks are variable even within one delivery to a site. Good clay bricks are generally free from cracks, have a good texture and are well burnt. Some of the main properties and their determination are given here.

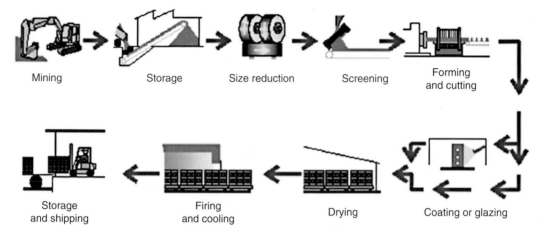

Figure 16.1 Brick manufacturing process. Reproduced by permission of the Brick Industry Association. (See the colour plate section for a full-colour version of this image.)

Appearance

Natural colours of clay bricks include red, white, yellow, brown and blue. The colour darkens as the temperature of firing increases. A light-coloured brick would suggest inadequate firing, and hence lower durability and strength, if the normal colour of this brick is known to be dark. Texture results from the method of forming or the surface treatment of green bricks.

Strength

Clay bricks may have strengths of up to $180\,N/mm^2$, but considerably lower strengths are adequate for the loads that are usual in small buildings. The Building Regulations require only $2.8\,N/mm^2$ for inner walls of two-storey houses. The compressive strengths of Class A and Class B engineering bricks are at least 70.0 and $50.0\,N/mm^2$, respectively.

Water Absorption

The absorption of water by bricks may be responsible for several problems that affect their durability. The limits for water absorption for Class A and Class B engineering bricks are 4.5% and 7%, respectively.

Frost Resistance

High strength and/or low water absorption are not satisfactory indices for frost resistance in bricks. There are three classes of frost resistance:

- **Class F:** These bricks are frost-resistant and should survive 100 cycles of freeze and thaw. These bricks can be used externally in situations of exposure to repeated cycles of freezing and thawing, for example, garden walls.
- **Class M:** These bricks are moderately frost-resistant and can be used in the construction of external walls of buildings.
- **Class O:** These bricks are not frost-resistant and should be used internally.

Thermal Conductivity

The thermal conductivity of a material depends on its density. Ordinary clay bricks have high density (typical: $2000\,kg/m^3$) and are poor insulators. In heat-loss calculations, the thermal conductivity of exposed brickwork is usually taken as 0.84 W/m K. This, however, varies with the moisture content of brickwork.

Fire Resistance

During the manufacturing process, clay bricks are fired at a temperature of approximately 1000 °C, which is higher than the temperature that may arise normally in fires in buildings. Therefore, clay bricks provide excellent resistance to fire.

Durability

The durability of bricks is much more likely to be a problem than is their strength. Durability problems are mainly associated with water absorption; therefore, it is important that bricks are selected according to the exposure conditions.

16.2.6 Deterioration of Brickwork

The durability of bricks depends on physical properties like density and porosity, and on the quality of the bricks used in the brickwork. The main cause of deterioration of brickwork is the absorption of water by the bricks.

Frost Damage

Frost damage is associated with freezing of water below the surface of the bricks. The bricks must be fairly porous in order to admit water. Water expands on freezing, and if the pores within the bricks cannot accommodate this expansion, the surface of the bricks cracks and crumbles away. This is known as **spalling** (Figure 16.2).

The brickwork should be protected from absorbing large amounts of water by providing a damp proof course (dpc), wherever applicable. The internal pore structure of bricks should be able to accommodate the expansion of water due to freezing.

Efflorescence

Some clay bricks may contain salts like sulphates of sodium and calcium. Porous bricks absorb water through their surfaces and the water dissolves these soluble salts. The salt solution moves to the

Figure 16.2 Brick spalling. (See the colour plate section for a full-colour version of this image.)

Figure 16.3 Brick efflorescence. (See the colour plate section for a full-colour version of this image.)

surface of the bricks as they start to dry out in the warmer/dry season. The water evaporates leaving a deposit of unsightly white powder, as shown in Figure 16.3. This is known as **efflorescence**.

The salt deposit on the surface does not cause any damage and it can be removed by brushing the surface of the brickwork. If the efflorescence is heavy and occurs below surfaces, crumbling of under-fired bricks may result.

Sulphate Attack

Sodium and calcium sulphates in clay bricks can also cause damage to the mortar, which eventually affects the brickwork. Porous bricks absorb water continually, and if there are sulphates in the bricks these will be dissolved. If the sulphate solution comes into contact with the mortar, it will react with the cement in the mortar. A new compound (calcium sulpho-aluminate) forms, which has more volume than the original compound (tricalcium aluminate). The mortar expands and causes damage to the brickwork.

Sulphate attack can be prevented by using bricks that have low soluble-salt content or by using sulphate-resisting Portland cement in the mortar.

16.2.7 Environmental Implications

An environmentally sustainable material produces only a small impact on the environment due to its manufacture and use in construction projects.

The raw material for clay bricks occurs in abundance; therefore, unlike other raw materials, a shortage of clay is not a concern. There is virtually no wastage of clay in brick manufacture, as any processed clay removed in the forming process before firing is recycled to the production stream. Transportation costs are reduced by siting the brick manufacturing plants near the sources of raw material. Bricks have a long life; this can be confirmed by checking the condition of the bricks used in buildings that are more than 100 years old. Even after demolition of buildings, the old bricks can either be used in construction again or can be crushed and used as aggregate or landscaping material.

16.2.8 COSHH

Brick manufacture involves quarrying of clay soil and its transportation to the manufacturing plant. These processes produce fine dust that contains respirable crystalline silica. The particles are so fine that they are deposited in the lungs when they are inhaled by workers. If this process continues for many years, crystalline silica can cause a lung disease known as silicosis. Silicosis takes many years to develop and can cause breathing problems, cough, sputum production and breathlessness. In severe cases, it can cause lung cancer and death. The HSE (Health and Safety Executive) has set a limit of $0.1\,mg/m^3$ for respirable crystalline silica. The prevention of silicosis can be aided with measures such as improved work practices and respiratory-protection equipment.

16.3 Aerated Concrete Blocks

Aerated concrete blocks are very light as compared to bricks, stone and dense concrete blocks. They contain numerous tiny air pockets, resulting in smaller mass and other desirable properties. Aerated concrete is a versatile material that is used in the construction of external walls, internal walls, foundations and suspended concrete floors.

16.3.1 Manufacture

The blocks are manufactured from sand, lime, pulverised fuel ash (PFA) and cement. Aluminium powder is added to the mix, which reacts with cement to form bubbles of hydrogen gas. The mix expands and hardens to form a cake, which is then cut into blocks. The next stage is to steam-cure the blocks, under high pressure, in an autoclave to produce a physically and chemically stable product. The blocks are light grey in colour and have a texture that is suitable for most types of plastering or rendering.

16.3.2 Size

Originally, only one or two types of block were available, but now a variety of blocks are manufactured, for example, standard, foundation and floor blocks. The standard size of blocks used in wall construction is $440 \times 215\,mm$ and they are available in several thicknesses, some of which are: 75, 100, 140, 150, 190, 200, 215, 260, 275 and 300 mm.

   For the inner leaf of cavity walls in dwelling houses, 100 mm thick blocks are used. For energy conservation and to satisfy the requirements of Building Regulations, blocks with a compressive strength of $2.9\,N/mm^2$ and thermal conductivity of $0.11\,W/m\,K$ are used. Trench blocks are specially made to construct the foundations of a building, and, similarly, floor blocks find their use in the construction of suspended concrete ground floors.

16.3.3 Properties

Appearance
The blocks are light grey in colour and have a rough texture that is suitable for most types of plastering or rendering.

Sound Insulation and Acoustic Control
A large number of air voids inside the blocks (microcellular structure) offers good sound and thermal insulation. The average sound-reduction index of 100 mm thick blocks with lightweight plaster is 40 decibels (dB). The sound-absorption coefficient depends on the frequency of sound; at a frequency of 500 Hz, the sound-absorption coefficient is 0.20.

Table 16.1

Property	Insulation blocks	Standard blocks	Foundation blocks
Compressive strength	$2.9\,N/mm^2$	$3.6\,N/mm^2$	$3.6, 7.3$ and $8.7\,N/mm^2$
Thermal conductivity	$0.11\,W/m\,K$	$0.15\,W/m\,K$	$0.15–0.20\,W/m\,K$
Average density	$460\,kg/m^3$	$600\,kg/m^3$	$600–750\,kg/m^3$
Vapour resistivity	60 MN s/gm	60 MN s/gm	–
Coefficient of linear expansion	$8 \times 10^{-6}/K$	$8 \times 10^{-6}/K$	–

Fire Resistance

Aerated concrete blocks are classified as non-combustible. Blocks that are 100 mm thick provide fire resistance of 2 hours and 4 hours for load-bearing and non-load-bearing elements, respectively. Blocks thicker than 100 mm will provide longer periods of fire resistance.

Durability

Aerated concrete blocks do not rot or decay and are resistant to freeze–thaw cycles. They have good resistance to sulphate attack.

Workability

Aerated concrete blocks are lightweight building units that can be safely and repetitively handled by a single person. They can be easily and accurately cut with woodworking tools, minimising the generation of solid waste.

Other Properties

The strength, thermal conductivity and other properties are summarised in Table 16.1. The blocks are classified according to their compressive strength. Manufacturers in the UK have devised special names for some of these blocks.

16.3.4 Environmental Implications

Aerated concrete blocks use up to 80% recycled materials, such as PFA, in their manufacture, and have played a major role in lowering the U-values of external walls of buildings. PFA, a stable material, is a by-product from coal-burning power stations, which would otherwise be used as landfill. Any waste during the manufacturing process is recycled into the next mix or used in other concrete products. Due to their light weight, which is partly due to the microcellular structure, less petrol/diesel is used in their transportation. Similarly, the microcellular structure of the blocks also results in remarkably high thermal insulation, which lowers energy consumption for the heating of buildings. After use, aerated concrete blocks can be broken up and recycled as aggregate.

16.4 Cement

Cement is one of the most important construction materials and is used in the production of concrete and mortar. The most commonly used type of cement is known as **Portland cement** (previously known as ordinary Portland cement) because the colour of solid cement resembles natural Portland stone.

16.4.1 Raw Materials

Portland cement is made from two raw materials: clay (or shale) and limestone (or chalk). Clay may be regarded as a mixture of three substances: silica (SiO_2), alumina (Al_2O_3) and iron oxide (Fe_2O_3).

Chalk and limestone are carbonates of calcium ($CaCO_3$). Both are natural materials: chalk occurs as a soft rock whereas limestone occurs as a hard rock.

16.4.2 Manufacture

Cement may be manufactured either by a wet process or by a dry process. The wet process involves crushing of chalk and mixing with clay slurry (clay–water mixture). The mixture is ground further into cement slurry and fed into a rotary kiln for firing.

The raw materials used in the dry process are clay (or shale) and limestone. The raw materials are crushed, mixed and then fed into a ball mill for further grinding. The powdered cement mixture is passed through a screen before feeding into the kiln.

The cement mixture is fed into the upper part of the rotary kiln (see Figure 16.4), which is maintained at a temperature of about 500 °C. Water is evaporated from the cement mixture, and the dry mixture is passed into the middle section of the kiln, which is maintained at a temperature of about 900 °C. Chalk/limestone is converted into lime at high temperature with the release of carbon dioxide gas.

$$\text{Chalk / limestone} (CaCO_3) \rightarrow \text{Lime} (CaO) + \text{Carbon dioxide} (CO_2)$$
$$(\text{Calcium carbonate}) \qquad\qquad (\text{Calcium oxide})$$

In the last stage of firing, the ingredients of the cement mixture, i.e. silica and lime, combine to produce nodules of cement, known as **cement clinker**. The cement clinker is cooled, mixed with about 5% gypsum to retard the setting process and ground to a fine powder.

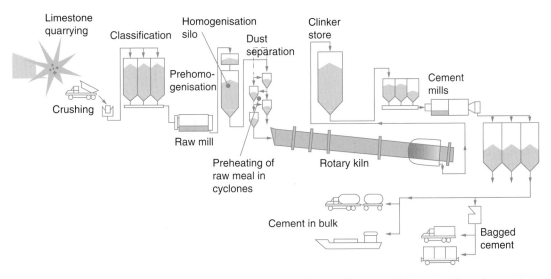

Figure 16.4 Cement manufacturing process. Reproduced by permission of Kääntee, U. (See the colour plate section for a full-colour version of this image.)

The wet process requires a large volume of water to make the clay–chalk slurry, which, in turn, consumes more energy than that of the dry process in evaporating the water. The dry process is, therefore, the preferred method where feasible.

16.4.3   Setting and Hardening of Cement

When cement and water are mixed, a chemical reaction known as **hydration** takes place, resulting in solid cement. During hydration, water and cement combine to form interlocking calcium silicate hydrate, which gives solid cement its hardness and strength. The heat evolved during the reaction is called the heat of hydration.

$$\text{Cement powder} + \text{water} \rightarrow \text{solid cement} + \text{heat}$$

Several factors can affect the rate of setting and hardening, temperature being one of them. An increase in temperature will increase the rate of the setting and hardening process. If the temperature falls below 0 °C, the setting and hardening process will stop. Similarly, the fineness (particle size) of cement affects the rate of setting and hardening. The rate of setting and hardening of cement increases with an increase in its fineness.

16.4.4   Constituents of Portland Cement

The raw materials of cement combine chemically in the kiln to produce cement that has properties completely different from their own. The four important constituents and their properties are summarised in Table 16.2.

16.4.5   Types of Cement

The classification of cements is based on their main constituents, for example, Portland–fly ash cement and Portland–pozzolana cement. The European Standard EN 197-1:2000 lists five types of cement that have a wide range of permitted constituents:

- CEM I: Portland cement;
- CEM II: Portland composite cement;
- CEM III: Blast-furnace cement;
- CEM IV: Pozzolanic cement;
- CEM V: Composite cement.

Table 16.2 Properties of the main constituents of cement.

Name	Chemical formula	% in OPC	Properties
Tricalcium silicate	$3CaO.SiO_2$	45	Short setting time; high early strength
Dicalcium silicate	$2CaO.SiO_2$	28	Long setting time; slow strength development but increases the durability of hardened cement
Tricalcium aluminate	$3CaO.Al_2O_3$	11	Quick setting (delayed by gypsum); attacked by sulphates
Tetracalcium aluminoferrite	$4CaO.Al_2O_3.Fe_2O_3$	9	This constituent has no contribution to setting or strength

Portland Cement CEM I

Formerly known as ordinary Portland cement (OPC), CEM I is manufactured to conform to British Standard BS EN 197-1:2000. It is the most commonly used cement in construction work throughout the world. The three strength classes of Portland cement, i.e. 32.5, 42.5 and 52.5, correspond to their lower characteristic strength in MPa (1 MPa = 1 N/mm^2) at 28 days. Where early strength of concrete is required, Portland cement 32.5R or 42.5R or 52.5R can be used. These cements are more finely ground than 32.5 N or 42.5 N or 52.5 N cements to enable faster hydration in the early stages.

CEM I cements are easy to procure, and concretes/mortars made using CEM I are versatile and durable. One major disadvantage is that, due to a high proportion of cement clinker, this is the least sustainable type of cement.

Factory-Made Composite Cements

Factory-made cements (CEM II, CEM III, CEM IV and CEM V) have Portland cement clinker varying from 5% to 94% (by mass) and one or more additional constituents. The additional constituents are selected from materials such as fly ash (PFA), blast-furnace slag, limestone and pozzolanic materials to reduce the environmental impact and promote sustainability. Some of these are described here; for more information, refer to BS EN 197-1:2000 and Lafarge Cement's publications.

Portland–Fly Ash Cement

Fly ash (PFA) is produced when coal is burnt in coal-fired power stations and furnaces. Fly ash has high silica and alumina content and is added to Portland cement to produce:

1) CEM II/A-V, which contains 6–20% siliceous fly ash.
2) CEM II/B-V, which contains 21–35% siliceous fly ash.

Calcareous fly ash may also be used to produce CEM II/A-W and CEM II/B-W cements. The use of fly ash in cement increases the strength and durability, improves sulphate resistance and reduces the heat of hydration, the risk of alkali–silica reaction and efflorescence. The use of fly ash has several environmental benefits as well, for example, recycling a waste by-product of coal-fired power stations.

Portland–Slag Cement

Blast-furnace slag is a by-product of iron production (see the subsections within Section 16.6.1). The slag is quenched in water to produce glassy granules that are similar to sand in appearance. Before mixing with Portland cement, the slag is dried and ground to a fineness that is similar to that of cement. There are two types of Portland–slag cement:

1) CEM II/A-S, which contains 6–20% slag.
2) CEM II/B-S, which contains 21–35% slag.

The use of blast-furnace slag in cement improves the sulphate resistance of concrete, workability and resistance to alkali–silica reaction. As with fly ash, the use of blast-furnace slag reduces the environmental impact by recycling the waste by-product of iron production.

16.4.6 Compressive Strength

The samples used for determining the compressive strength measure 40 × 40 × 160 mm, and are made from 1:3 cement–sand mortar with a water–cement ratio of 0.5. The sand used is a standard CEN sand of specified grading with particle size between 1.6 mm and 80 μm (BS EN 196-1). The samples

Table 16.3

Strength class	Compressive strength (MPa)			Maximum strength (MPa)
	2 day	7 day	28 day	28 day
32.5 N	–	≥ 16.0	≥ 32.5	52.5
32.5 R	≥ 10	–	≥ 32.5	52.5
42.5 N	≥ 10	–	≥ 42.5	62.5
42.5 R	≥ 20	–	≥ 42.5	62.5
52.5 N	≥ 20	–	≥ 52.5	–
52.5 R	≥ 30	–	≥ 52.5	–

(Extract from BS EN 197-1: 2000)

are first tested to determine their flexural strength and then broken (if not already) for use in the compression test. A letter is added after the strength: 'N' for normal, 'L' for low and 'R' for rapid rates of strength development. Table 16.3 shows the strength classes of Portland cement.

16.4.7 Environmental Implications

Cement manufacturing involves the quarrying of raw materials, their transportation to the cement works, firing the raw materials in a kiln and delivery of the finished product. Each stage of cement manufacture has some impact on the environment. Extraction of limestone from quarries by blasting produces dust, gases, noise and damage to the landscape. The emission of dust and exhaust gases can be reduced with the use of suitable equipment. The quarrying sites are reintegrated into the countryside after extraction of the raw materials. Cement manufacturing requires large amounts of energy to heat the kilns and it is responsible for producing large amounts of carbon dioxide (CO_2) emissions. CO_2 is released into the atmosphere if fossil fuels are used to generate the energy. CO_2 is also released directly into the atmosphere when chalk or limestone is heated in the kiln, producing lime and CO_2. Cement manufacturers use electrostatic precipitators and other measures to reduce the dust created by the kilns.

CEM II cements use by-products like fly ash, blast-furnace slag etc., resulting in fewer CO_2 emissions and reduced demand for raw materials. Also, the slag and fly ash that would otherwise go to landfill are put to a better use.

16.4.8 COSHH

Cement is highly alkaline and it is important for workers to avoid contact with cement or concrete. Contact with the skin may cause irritation, burns and dermatitis. Cement contains chromium (VI) which may cause allergic reactions.

16.5 Concrete

Concrete is used in the construction of many building elements, and hence is an important building material. It is a composite material made by mixing together cement, aggregates (sand and/or gravel) and water. The proportion of these ingredients affects not just the strength but also the ease with which concrete can be placed and compacted. The function of cement is to act as a binder, i.e. to glue the aggregates together. Cement mixes with water to form a paste that coats all aggregates in the mixing process. Although cement is responsible for the strength of concrete, it is not used without the aggregates for the following reasons:

- Cement is the most expensive ingredient of concrete. Its volume in concrete is kept to a minimum, as per the mix design.
- Cement shrinks after setting and hardening. This may happen in concretes in which the amount of cement used is far more than what is required. Aggregates help to reduce the cracking in concrete.
- Properly made concrete is likely to be stronger than neat cement.

16.5.1 Raw Materials

The raw materials of concrete are:

- Cement, which acts as a binder;
- Fine aggregate (sand);
- Coarse aggregate (crushed or uncrushed gravel);
- Water.

16.5.2 Manufacture of Concrete

Concrete is made by mixing the raw materials in the correct proportions. A chemical reaction is initiated when cement comes into contact with water. This is known as **hydration**, and is responsible for the setting and hardening of cement. All constituents of cement undergo hydration and contribute to the final product. Most of the early strength of concrete is due to tricalcium silicate, whereas dicalcium silicate, which reacts slowly, is responsible for the development of strength at a later stage.

Concrete is placed in formwork to take the desired shape of the element/component being cast and is vibrated to achieve adequate compaction. **Compaction** of concrete is necessary to remove any air voids formed during mixing and placing.

Cement needs water for setting and hardening; therefore, concrete is never allowed to dry out during the first 2 to 3 weeks. This is known as **curing** of concrete.

16.5.3 Concrete Mix

The proportions of aggregates, cement and water for a particular mix depend on the part of the building where it is to be used. It is therefore important to mix the constituents of concrete in the correct proportions so that the final product has the required workability and strength. A typical concrete mix for a strip foundation of a two-storey house is 1 part cement, 3 parts fine aggregate and 6 parts coarse aggregate; this is written in a shorter form as **1:3:6** concrete. Concrete can also be made by using all-in aggregate (ballast) which is a mixture of fine and coarse aggregates. If we use all-in aggregate, then the equivalent concrete mix is 1:7 (approximately). The quantity of water can be determined from the given water–cement (w/c) ratio.

There are two methods of determining the quantities of the constituents of concrete:

1) Mixing by volume.
2) Mixing by mass.

Mixing by Volume

The proportions of cement, fine aggregate and coarse aggregate are measured by volume, and the mix thus obtained is known as a **nominal mix**. Typical nominal mixes are:

- **Mass concrete:** Mix 1:3:6, for use in narrow strip and deep strip foundations of dwelling houses.
- **Reinforced concrete:** Mix 1:2:4, 1:1½:3 and 1:1:2, for use in slabs, beams, columns and foundations of buildings and other structures.

In nominal mixes, the ratio of fine aggregate to coarse aggregate is always 1:2. This will ensure that the volume of fine aggregate is enough to fill the voids between the coarse aggregate and hence result in a dense concrete mix.

Mixing by volume results in poor quality control as the water absorbed by the aggregates is not taken into account.

Mixing by Mass

The proportions of the constituents of concrete are measured by dry mass. The mixes so obtained are known as prescribed mixes, standard mixes, designated mixes and designed mixes. Details of these are given in BS 5328; a summary is produced here.

Prescribed Mixes

The mix is specified by its constituent materials and the quantities of those constituents to produce a concrete with the required performance. The person requiring the concrete mix must specify the nominal maximum size of aggregates, permitted types of aggregate, the mix proportion by weight and the workability.

Standard Mixes

Depending on the characteristic compressive strength of concrete, five types of standard mix can be produced. ST1 is the weakest at $7.5\,N/mm^2$ and ST5 the strongest at $25\,N/mm^2$. The mix is first selected from BS 5328 and reference is made to other tables for selecting the materials and the mix proportions.

Designated Mixes

Designated mixes are specified by considering the site conditions and then selecting the application (from BS 5328) for which the concrete is to be used. The person requiring the concrete mix specifies:

- Whether the concrete is to be reinforced or unreinforced;
- The aggregate size, if it is not 20 mm;
- Whether the concrete is to be exposed to a chloride-bearing environment, and/or to severe freezing conditions while wet;
- The workability.

Designed Mixes

The mix is specified in terms of a strength grade: either compressive strength grades (C7.5 to C60) or flexural strength grades F3 to F5. The specified strength grade is subject to restriction on the maximum or minimum cement content, maximum free water–cement (w/c) ratio, nominal maximum size of aggregate and any other properties required. An important part of designed mixes is to perform strength testing to check conformity to the specifications.

16.5.4   Properties of Fresh Concrete

Workability

The quantity of water used in making a concrete mix is perhaps the most important factor in influencing the workability of fresh concrete and the strength of matured concrete. The amount of water in a mix is expressed as the w/c ratio. For example, if a concrete mix is made by mixing 50 kg of cement, 300 kg of all-in aggregate and 30 kg of water, then the w/c ratio is:

$$\text{w / c ratio} = \frac{\text{Amount of water}}{\text{Amount of cement}} = \frac{30\,\text{kg}}{50\,\text{kg}} = \mathbf{0.6}$$

The workability of concrete is basically the ease with which it can be placed and compacted, and is proportional to the w/c ratio. The typical range of the w/c ratio is from 0.4 to 0.8. When the w/c ratio is low (for example, 0.3), the mix is dry and difficult to compact. By increasing the amount of water in concrete, compaction becomes much easier. Such a mix will have a high w/c ratio and high workability. The strength of solid concrete, however, is inversely proportional to the w/c ratio; highest strength could be achieved at a w/c ratio between 0.3 and 0.4. A large amount of water will produce a highly workable mix, but leaves voids in the solid concrete on evaporation. This will result in a lower strength of concrete. Other factors affecting workability are the shape, size and texture of the aggregates. Rounded aggregates produce greater workability whereas angular aggregates produce the lowest levels of workability. As the roughness of the aggregates increases, the workability of the mix decreases. The larger the maximum allowable size of aggregate, the more workable the concrete, because the surface area of bigger aggregates is less than that of smaller aggregates; the cement paste has to coat a smaller area, resulting in more workable concrete.

Tests for Workability

The workability of concrete may be determined by performing a slump test, compaction factor test or VB consistometer test.

The slump test is the most commonly used site and laboratory method for determining the workability of concrete. The test consists of filling an open-ended conical mould with wet concrete in four equal layers, each being tamped 25 times with a tamping rod. The top of the concrete is levelled off and the cone removed. The drop in the level of concrete, i.e. the slump, is measured, as shown in Figure 16.5(a). A concrete mix of very low w/c ratio may produce zero slump. On the other hand, a concrete mix with a large amount of water may produce a collapse slump. The result must match the specification used in the mix design.

The compaction factor test gives a more accurate measure of the workability of fresh concrete. The upper hopper of the apparatus, shown in Figure 16.5(b), is filled with concrete. The flap at the bottom of the hopper is opened, letting the concrete fall into the lower hopper and some compaction takes place due to this action. Next, the flap of the lower hopper is opened to let the concrete fall into a compacting cylinder whose mass is already known. Any excess concrete is removed from the top of the cylinder and the mass of this partially compacted concrete is found. The next step is to compact the concrete with a vibrator, adding more concrete until the compacting cylinder is full of compacted concrete. The mass of fully compacted concrete is found and the compaction factor determined from the following formula:

$$\text{Compaction factor} = \frac{\text{Mass of partially compacted concrete}}{\text{Mass of fully compacted concrete}}$$

(a)

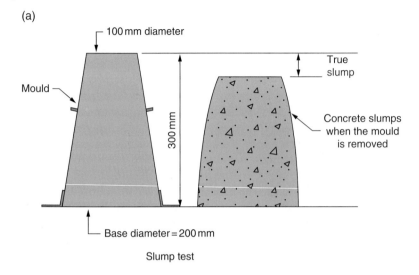

Slump test

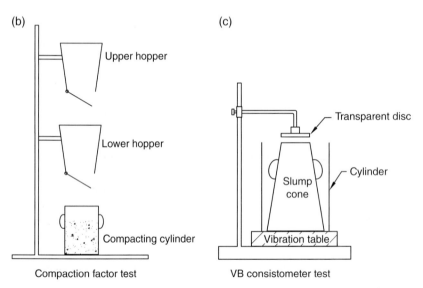

Figure 16.5

A compaction factor of 0.95 and above indicates high workability; a compaction factor of less than 0.85 indicates low workability.

The VB consistometer test is used for concrete that has low workability. The slump cone (Figure 16.5(c)) is filled with concrete in four layers, tamping each layer 25 times. The slump cone is removed and the transparent disc lowered so that it just touches the top of the slumped concrete. The concrete is vibrated and the time taken to achieve full compaction is recorded. At full compaction, the transparent disc becomes completely coated with concrete and the time taken for full compaction, known as the VB time, is a measure of the workability of concrete.

Time in seconds	Workability
0 to 1	High
1 to 25	Medium
Over 25	Low

16.5.5   Properties of Hardened Concrete

Density

The density of concrete depends on the type of aggregate used. If dense aggregates are used, then the density will be $2400\,kg/m^3$. The use of lightweight aggregate produces concrete with a density less than $2000\,kg/m^3$.

Modulus of Elasticity

The modulus of elasticity of dense concrete ranges between 20700 and $34500\,N/mm^2$.

Thermal Conductivity

Concrete made from lightweight aggregate provides better thermal insulation: its thermal conductivity is approximately $0.6\,W/m\,K$. The thermal conductivity of dense concrete is $1.9\,W/m\,K$.

Strength of Concrete

Concrete is very strong in compression but weak in tension. The strength of concrete is affected by several factors, which are explained below:

- **Voids:** The space occupied by coarse aggregate is the sum of the volume of solids and the volume of air voids. In order to obtain dense concrete, the voids must be filled with fine aggregate, which is only possible by proper compaction. Insufficient compaction may not achieve this, resulting in weak concrete.
- **Water–cement ratio:** The most important factor that affects the strength of concrete is its w/c ratio. If the volume of water used to prepare the mix is more than the optimal requirement, then the excess water remains free while the concrete is still in a semi-fluid state. As the concrete becomes solid, the excess water evaporates leaving voids which make the concrete weaker.
- **Fine aggregate:** There must be an adequate quantity of fine aggregate in the mix to fill the voids between the coarse aggregate. An insufficient quantity of fine aggregate (a very low fine–coarse aggregate ratio) causes air voids to form, resulting in weaker concrete.
- **Shape and texture of aggregates:** The shape and texture of aggregates also affect the strength of concrete. The crushed aggregates interlock with one another and produce a higher strength than uncrushed aggregates.

Durability of Concrete

It is important that the constituent materials of concrete are selected carefully. If concrete is designed, mixed and compacted properly, then it can easily last for several decades. Concrete should be impervious; porous concrete is susceptible to damage from frost and chemicals, as explained in Section 16.5.6.

Fire Resistance

Concrete is a non-combustible material and reasonably stable up to about 550 °C. Because of the low thermal conductivity of concrete, heat is transferred at a very slow rate. It cannot be set on fire and does not emit any toxic fumes if subjected to high temperatures.

16.5.6 Deterioration of Concrete

The durability of concrete can be affected by sulphate attack, frost attack, alkali–silica reaction and corrosion of its steel reinforcement.

Frost Attack

Frost can affect fresh, as well as mature, concrete. If concreting is done in freezing conditions, the water in the mix can freeze and affect the setting and hardening process. Also, water expands on freezing, which can cause damage to concrete. This can be avoided by:

- Protecting the concrete from frost by covering the freshly placed concrete with insulation;
- Using hot water to accelerate the setting time;
- Using rapid-hardening Portland cement.

Mature concrete is also likely to be attacked by frost if it is porous. Water is absorbed through pores in the concrete, which expands on freezing, causing damage. Cracking and spalling of the surface of concrete occur due to frost attack. It is important to remember that repeated freeze–thaw cycles are more damaging than steady freezing.

Frost attacks on porous concrete can be avoided by protecting the concrete with a layer of suitable impermeable material.

Sulphate Attack

Sulphates in soils and in sea water react with the tricalcium aluminate content of cement to form tricalcium sulpho-aluminate. Tricalcium sulpho-aluminate occupies a greater volume than tricalcium aluminate. This has a disruptive effect on concrete, causing cracking and spalling at the surface.

Sulphate attack can be prevented by replacing Portland cement with sulphate-resisting cement in the concrete mix.

Alkali–Silica Reaction

Alkali–silica reaction (ASR), also known as concrete cancer, is a chemical reaction between the alkalis in Portland cement and the reactive forms of silica that are present in some aggregates. Cement (and concrete) contains oxides of sodium, potassium and calcium, which react with water to produce a highly alkaline (pH > 13) pore solution of concrete. Silica, which is a constituent of the aggregates, is acidic and, in the presence of water, reacts with the alkalis to produce viscous alkali silicate gel. The gel forms around, and within, the aggregate and expands on absorbing more water from the cement paste. Eventually, the gel may crack the aggregate and the surrounding concrete when the swelling pressure exceeds the strength of the aggregate.

It is quite difficult, if not impossible, to minimise the ASR once it has begun. Although costly, the use of lithium has been found to be effective. The best strategy to minimise the ASR is to prevent its occurrence by using appropriate materials in the concrete mix, for example:

- Use cement with a low alkali content;
- Avoid aggregate that has a reactive form of silica;
- Use one or more supplementary cementitious materials, such as PFA and blast-furnace slag, to reduce the amount of alkali in the mix.

Corrosion of Steel Reinforcement

Steel bars are used in reinforced concrete to resist the tensile stress that plain concrete cannot resist. The steel is protected from corrosion by the alkaline environment of the concrete. If the alkaline environment is neutralised and the concrete is permeable, water can seep through and cause corrosion of the steel. Corroded steel will lose the bond with the concrete that is necessary to transfer tensile stress from the concrete to the steel.

To protect the steel reinforcement, the concrete must be impermeable to water and must also be free from substances that can chemically attack the steel.

16.5.7 Environmental Implications

The main materials required to produce plain concrete are cement and aggregates. Similarly, the main materials required for reinforced concrete are plain concrete and steel. Production of cement and steel requires a number of raw materials, i.e. limestone/chalk, clay/shale, iron ore and coke. These materials, which are extracted from quarries/mines, have a major impact on the environment, as explained in Section 16.4.7 (cement) and within the subsections of Section 16.6.1 (steel). The manufacture of steel produces large volumes of waste water, some of which is recycled.

The huge impact that concrete has on the environment can be reduced by replacing Portland cement with CEM II or other cements. For example, Portland–fly ash cement (CEM II/B-W) saves about 35% in limestone and clay.

The impact on the environment can also be reduced if concrete is crushed and recycled as aggregate for new concrete. It must be free from contaminants like plaster and wood. Recycling old concrete saves landfill space and reduces the need for gravel quarries.

16.6 Metals

Metals can be classified into two groups: ferrous and non-ferrous. Ferrous metals contain iron, typical examples being cast iron, wrought iron and the various types of steel. Non-ferrous metals do not have any iron; typical examples of non-ferrous metals used in buildings are copper, lead, aluminium and zinc. Ferrous metals, mild steel in particular, are used extensively in the construction of buildings, dams, tunnels, bridges and other structures.

16.6.1 Ferrous Metals

Ferrous metals and alloys, such as pure iron, cast iron, wrought iron, mild steel etc., contain iron as the parent metal. The word ferrous comes from **ferrum**, which is the Latin name of iron (chemical symbol: **Fe**). In nature, iron is found in minerals formed by its combination with other elements, for example, iron and oxygen combine to produce iron oxides. **Haematite** (Fe_2O_3) and **magnetite** (Fe_3O_4), which are oxides of iron, are the main ores of iron and are mined commercially.

Raw Materials

The raw materials used to produce pig iron in a blast furnace are iron ore, coke and limestone. The iron ore is extracted from the ground and partially refined to remove some of the impurities. Coke is made by heating coal; limestone is a carbonate of calcium and occurs naturally. The iron ore is finely divided and roasted with coke and limestone to remove a large amount of the impurities that occur naturally in the ore.

Manufacturing Process

The iron ore is heated in a blast furnace with limestone and coke (impure carbon), where several chemical reactions take place, resulting in molten iron:

$$C + O_2 \rightarrow CO_2 \ (C - \text{carbon}; \ O_2 - \text{oxygen})$$

$$CO_2 + C \rightarrow 2CO \ (CO_2 - \text{carbon dioxide}; CO - \text{carbon monoxide})$$

Carbon monoxide acts as a reducing agent and reacts with the iron ore to give molten iron, which is collected at the bottom of the furnace (Figure 16.6).

$$Fe_2O_3 + 3CO \rightarrow 2Fe + 3CO_2$$

The limestone in the furnace decomposes, forming calcium oxide (CaO). Calcium oxide combines with impurities to make slag ($CaSiO_3$), which floats on top of the molten iron and can be removed.

$$CaO + SiO_2 \rightarrow CaSiO_3$$

The process results in pig iron and slag. Slag is used in cement manufacture and as an aggregate in road construction and other applications. Pig iron contains 4–5% of carbon and other impurities like manganese, silicon, phosphorus and sulphur, which can be removed to give different types of steel.

One of the biggest drawbacks of blast furnaces is the production of carbon dioxide gas, which is a greenhouse gas. New iron-making processes have been developed, which reduce the emission of CO_2 gas.

All ferrous metals contain carbon. This is due to the use of coke in the blast furnace during the smelting of the ore. Variations in the carbon content of ferrous metals have important influences on their properties. In general, an increase in the carbon content causes a reduction in ductility and ease

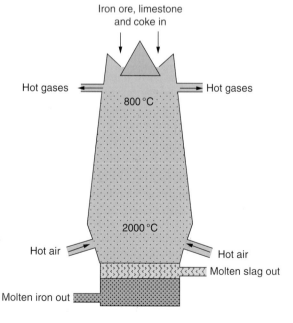

Figure 16.6 Blast furnace.

of welding, but an increase in the hardness. In steel, tensile strength increases with an increase in the carbon content up to about 1.5%.

Iron and Its Alloys
Cast iron
Cast iron is made by re-melting pig iron together with steel and cast iron scrap. Its high carbon content (2.5–4.5%) makes it free-running and suitable for castings. The tensile strength varies from 155 to 400 N/mm^2. Cast iron was used in the past for drainpipes, guttering and water tanks.

Wrought Iron
Wrought iron is made by melting pig iron with iron oxide. It is very soft as it contains a very small proportion of carbon, i.e. 0.02–0.03%. Wrought iron is moderately strong in tension, failing at about 355 N/mm^2, but is very tough and resistant to impact loads.

Steel
Steel is produced by removing impurities, such as sulphur and phosphorus, from pig iron and adjusting the carbon content. There are many types of steel:

- **Low carbon steel** contains up to 0.15% carbon. It is soft, malleable and ductile, and therefore used for wiring and sheets.
- **Mild steel** contains 0.15% to 0.25% carbon and is strong in tension, ductile and suitable for rolling into structural sections, strips and sheets.
- **Medium carbon steel** contains 0.25% to 0.50% carbon and is suitable for making nuts and bolts.
- **High carbon steels** contain 0.5% to 1.5 % carbon. The carbon content makes these steels both tough and brittle. These are used for making cutting tools, chisels and files.

Properties of Steel
Malleability and Ductility
Malleability is the property of a metal by virtue of which it can be deformed by compression without cracking or rupturing. Ductility is the ability to deform plastically without fracture under a tensile force. A material is malleable if it is possible to roll it into sheets, and ductile if it can be drawn into wires. On stretching ductile metals, we can see clearly the formation of necking (see Section 9.6.2) before failure, whereas in brittle materials there is no necking. This property (ductility) allows materials to redistribute stresses that may build up at weak points.

Mild steel is both malleable and ductile; it is used in manufacturing tools, tubes, pipes, structural sections, reinforcement bars and other applications.

Hardness
The resistance of a material to indentation is known as hardness. The term can also apply to deformation from scratching and cutting. In engineering, the Brinell test is usually performed to determine the hardness of a metal. A hardened metal sphere is placed on the metal to be tested and a force applied. The load divided by the surface area of the indentation on the test sample gives the Brinell hardness number. The hardness of mild steel is 120 HB.

Density
The density of steel is 7850 kg/m^3. This shows that steel is far denser than common building materials like bricks, blocks, cement, concrete and timber.

Thermal Conductivity

Steel, like other metals, is a good conductor of heat, or a poor insulator. Its coefficient of thermal conductivity (λ-value) is $50\,W/m\,K$, which means that the heat loss through steel is about 60 times that in bricks.

Thermal Movement

The coefficient of linear expansion of steel is $11-13 \times 10^{-6}/°C$.

Modulus of Elasticity

The value of the modulus of elasticity for steel is $210\,kN/mm^2$. Young's modulus is a measure of the stiffness of a material, or its resistance to compression or extension. For comparison, the modulus of elasticity of copper and aluminium is $117\,kN/mm^2$ and $70\,kN/mm^2$, respectively. It can be determined by performing a tensile test on the given metal using a piece of apparatus called a **tensometer**.

Electrical Conductivity

Steel has a low electrical resistance, or is a good conductor of electricity. Its electrical conductivity is $1.1 \times 10^7/\Omega m$.

Corrosion of Iron/Steel

Iron is made from its ores by a reduction process (removal of oxygen). As soon as iron/steel components are manufactured, their corrosion begins if suitable protection is not provided. Corrosion of iron/steel is caused by the following processes:

- Oxidation;
- Electrolytic action.

Oxidation of iron/steel, also known as rusting, causes a lot of damage to unprotected components. The process involves a reaction between oxygen and iron in the presence of water to form ferrous oxide (iron oxide) Fe_2O_3. $2H_2O$ or ferric hydroxide $Fe\,(OH)_3$. These products are reddish-brown in colour, and commonly known as **rust**. Once rusting starts, the porous oxides formed will allow more air and water to enter, causing more damage. Both air and water are required for rust to form and, in the absence of one of these, rusting will not occur. This can be shown by performing a simple experiment in which iron nails are put into test tubes and exposed to different types of environments (see Section 3.2.1 for details).

When two dissimilar metals are in contact, **electrolytic corrosion** occurs in the presence of water. This is explained in Chapter 3.

Protection of Steel from Corrosion

Steel can be protected from corrosion by painting, galvanising or other treatments. Refer to Sections 3.4.1 and 3.4.2 for details.

Environmental Implications

Steel is one of the main materials used in construction; about 30% of its total production is used in construction. Steel production involves several processes including the mining/quarrying of raw materials, the production of coke from coal, and the smelting of iron ore in a blast furnace. The mining of iron ore results in some destruction of the landscape, production of dust and gases and noise from the mining/transportation of raw materials. When mining is over, the sites are used as landfills, which can become a problem if toxic substances are dumped on such sites.

Iron-making requires large amounts of coal to produce coke, which is an important material in making iron by reduction of the iron ore. The coking process emits particulate matter, volatile organic compounds (VOCs), ammonia, methane, phenol, carbon monoxide and sulphur oxides. Most of these by-products are toxic but commercially very useful for many other industrial processes.

About 400 litres of waste water are generated during the production of one tonne of coke. Water is required for the cooling of the coke oven gas and the processing of ammonia, naphthalene, phenol and light oil. This is only a small fraction of the total water requirement; about 350000 litres of water are required to produce one tonne of steel. The steel-making sector is a very large consumer of energy, and therefore a major contributor to greenhouse gas emission.

Iron and steel are the most recycled materials, as it is cheaper to recycle them than to mine iron ore. Recycling reduces piles of old metal and all the forms of pollution that are associated with the mining of iron ore. Since the scale of mining is reduced, recycling saves a lot of energy as well.

16.6.2 Non-Ferrous Metal: Aluminium

Non-ferrous metals, such as aluminium, copper, brass, lead and zinc, are used in construction, but in this section only aluminium will be described. Aluminium is a lightweight, malleable, ductile and durable metal that has numerous uses in the construction industry. It is a sustainable material as the physical properties of the metal after recycling are the same as those of the original metal.

Raw Materials
Aluminium occurs in the Earth's crust as hydrated aluminium silicates and oxides; the usual aluminium ore is **bauxite**, which is basically impure aluminium oxide. The traditional smelting methods cannot be used to extract aluminium from bauxite because a very high temperature is needed. Instead, aluminium is extracted from its ore by electrolysis, the details of which are given in Chapter 3. The chemical symbol of aluminium is **Al**.

Properties
The important properties of aluminium include its attractive appearance, light weight, good thermal and electrical conductivities and high resistance to corrosion. Pure aluminium is comparatively soft and only has moderate tensile strength. However, its alloys with other metals (silicon, copper or magnesium) have considerable strength, making them suitable for use in structural members, scaffolding and ladders. Aluminium is a ductile metal, enabling it to be drawn into wires or extruded to structural members, tubes, pipes, gutters and door/window frames. When exposed, a thin layer of aluminium oxide builds up on the surface of fresh aluminium, which protects it from corrosion. This layer can be strengthened further by anodising the aluminium.

Aluminium has a high electrical conductivity and a high thermal conductivity.

Hardness
Pure aluminium is very soft, its hardness being 15 HB. When aluminium is alloyed with other metals, its hardness increases and ranges between 60 and 100 HB.

Fire Resistance
As with steel, the fire resistance of aluminium is not very high. Its melting point is 660 °C.

Durability
Aluminium is a durable metal with high resistance to corrosion. When exposed to the atmosphere, a hard layer of oxides forms on the surface which protects the metal from corrosion.

The properties of aluminium are compared with those of mild steel and copper in Table 16.4.

Table 16.4 Comparison of some properties of mild steel, aluminium and copper.

Metal	Density (kg/m^3)	Modulus of elasticity (kN/mm^2)	Coefficient of linear expansion (per °C)	Thermal conductivity (W/m K)	Electrical conductivity (siemens/m)	Tensile strength (N/mm^2)
Mild steel	7850	210	$11–13 \times 10^{-6}$	50–80	1.1×10^7	400
Aluminium	2700	70 (alloys)	23.5×10^{-6}	234	3.7×10^7	100 (pure)
						250 (alloy)
Copper	8900	117	16.6×10^{-6}	400	5.85×10^7	250

Note: The tensile strength of a metal depends on its composition

Environmental Implications

Bauxite is usually mined by opencast mining and, as with any other mining operation, the landscape and the habitat are destroyed. In opencast mining, the top layer of soil is removed to reach the ore underneath. If the top layer of soil is not reinstated after the mining operation is over, vegetation may not return.

Aluminium can be extracted only by electrolysis, which uses a huge amount of electricity. If fuels are burnt to generate electricity, the carbon dioxide emissions into the atmosphere cause pollution and global warming.

Nowadays, scrap aluminium is collected and recycled to save a large amount of electricity and protect the environment from various types of pollution. The recycling process consumes only about 5% of the energy used to extract the metal from bauxite ore.

16.7 Timber

Timber is one of the main materials used in the construction of low-rise domestic buildings. Its thermal properties, lightness and strength have, over time, encouraged builders to devise timber-framed house construction in countries with a temperate climate. Timber is an organic material obtained from trees. When a tree reaches maturity, it is felled, and the wood, obtained mainly from the trunk, is cut to suitable sizes. The moisture content of raw wood is reduced by a process known as **seasoning**. After seasoning, the wood is known as timber and is ready for use as a construction material. Timber is divided into two classes: softwood and hardwood. Softwood is obtained from evergreen trees, whereas hardwood is obtained from deciduous trees. Deciduous trees have broad leaves which are shed in the autumn. All softwoods are not soft and, similarly, all hardwoods are not hard.

A tree consists of three parts: the roots, the trunk and the crown. The roots spread out in the soil and absorb water and minerals; the solution of these is known as **sap**. The trunk, which eventually becomes the source of timber, supports the tree and transports the sap to the crown through the sapwood. The leaves, which make the crown, are responsible for producing the sugars and starches needed for the growth of the tree. Leaves absorb carbon dioxide from the atmosphere and, in the presence of sunlight and chlorophyll (the green substance in the leaves), convert the sap into sugars and starches. Wood tissue is composed of various types of cells, the walls of which are made of cellulose and lignin. Cellulose $(C_6H_{10}O_5)_n$ is a long-chain polymeric polysaccharide carbohydrate that forms the primary structural component of green plants. The lignin binds together the cellulose material to give the wood its strength and rigidity.

Figure 16.7 Cross-section of a tree trunk.

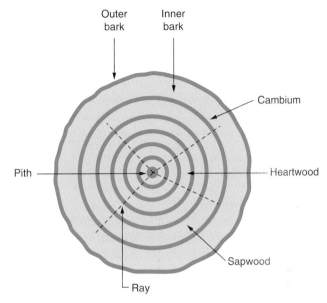

Figure 16.7 shows a horizontal section of a tree trunk, and illustrates the parts involved in the growth process of a tree.

The trunk and the branches of the tree are covered with **outer bark**, which protects the tree. It is formed from the dead inner bark cells.

The **inner bark**, found between the outer bark and the cambium, acts as a food supply line by carrying sap from the leaves to the rest of the tree.

The **cambium** is a very thin layer of growing tissue that produces new cells. Each year a new layer of cells (**sapwood**) is produced, making the cambium layer move further outwards from the centre of the trunk. This layer of sapwood, also known as an annular ring, is formed during spring and summer.

As new sapwood layers are formed, the cells in older sapwood layers become inactive and die to form **heartwood**. Heartwood is denser and more durable than sapwood. The central core of the tree is known as **pith**. It decays as the tree becomes older and has no commercial value.

16.7.1 Seasoning

Raw timber is not suitable for immediate use in buildings as it contains a lot of moisture. Seasoning is the process of removing a certain proportion of water from raw timber to make it suitable for use in buildings. Seasoned timber is lighter, stronger, easier to work with and easier to finish with paint/varnish. Seasoning may be natural (air seasoning) or artificial (kiln seasoning).

Air Seasoning

Timber is protected from sun and rain, and stacked off the ground to allow free circulation of air. The drying process should be slow to protect the timber from excessive shrinkage. Winter is the best time to start air seasoning, when the air temperature is low and the relative humidity of air is high, to ensure slow drying.

Kiln Seasoning

Air seasoning may take several months, whereas in kiln seasoning, an external source of heat energy is used to accelerate the drying process. The timber is stacked inside a chamber in which the temperature and humidity can be controlled to produce the best results. Each species requires a different optimum temperature and humidity to produce the desired moisture content.

16.7.2 Properties

Appearance

Timber is available in a variety of colours such as white, light yellow, orange, light brown, dark brown and others. Different species of tree produce different timber colours, which can change with age and by the application of finishes. The texture of a timber depends on the size and arrangement of its cells. Generally, softwoods have fine textures but hardwoods may have fine or coarse textures. The wood grain can also enhance the appearance of timber.

Density

Timber is a light material with the density of different species ranging from 160 to $1250 \, \text{kg/m}^3$. Although all species of tree are composed of the same substances, the differences in their microstructure affect the values of timber density. The densities of the most commonly used softwoods in the UK range from 380 to $550 \, \text{kg/m}^3$.

Thermal Conductivity (λ)

The thermal conductivity of timber depends on its moisture content and density; lightweight timber has better thermal insulation than dense timber. There are numerous air pockets within the cellular structure of timber that make it a natural insulator. The thermal conductivity values of timber used in buildings range between 0.08 and 0.18 W/m K (approximately).

Acoustic Properties

The cellular structure of timber has minute interlocking pores which convert sound energy into heat energy by frictional/viscous resistance, achieving sound insulation. The porous structure of timber is also responsible for the absorption of sound. Typical values of the sound absorption coefficient for wood blocks on a solid floor are: 0.05 at 500 Hz and 0.1 at 2000 Hz.

Strength

Timber is a lightweight but strong material with a high strength–weight ratio both in tension and compression. There is a large variation in the strength as there are so many different species of tree producing timber. Generally, the strength of timber is directly proportional to its density, and inversely proportional to its moisture content. Defects such as knots can also affect timber strength. The tensile strength of timber ranges between 2.5 and $14.0 \, \text{N/mm}^2$ and the modulus of elasticity ranges between 4500 and $20000 \, \text{N/mm}^2$.

Fire Resistance

In the case of a fire, the temperature of timber components does not rise quickly, as the material is a poor conductor of heat. Timber will ignite at a temperature of about 220–300 °C, but the charring of the timber surface acts as an insulator and provides protection to the inner parts of the components.

Hardness

Hardness is an important property to consider when a flooring material is selected. Flooring made of harder timber lasts longer and requires less maintenance than softwood flooring.

16.7.3 Deterioration

Timber is a durable material but can deteriorate if not protected from sunlight, chemicals, fungi, insects and other damage-causing factors. Because timber is basically cellulose, it is used as a source of food by fungi (plants) and insects. Details of these causes of deterioration and the protection of timber are discussed here.

Excessive Loading

If the load applied to structural elements is less than the design load, the stresses developed will be within permissible limits. If, for some reason, the applied loading exceeds the design load, then the stresses developed will be greater than the permissible stresses and may cause unacceptable bending and possibly lead to fracture.

Sunlight and Rain

The ultraviolet radiation in sunlight causes discoloration; excessive rain causes cracks and growth of moulds and algae.

Fire

Timber is a combustible material and decomposes at high temperatures to produce charcoal and combustible gases such as CO and methane. The propagation of fire from the surface of a component to the inside is slow because of the low thermal conductivity of timber. The charred surface of the timber component also acts as an insulator and protects the interior of the section.

Fungal Attack

Fungi are plants without leaves and branches, which feed on the cellulose of timber. For their growth, they require moisture, oxygen and a temperature between 20 °C and 30 °C.

The spores (seeds) of fungi are dispersed through the air. They settle on timber to produce fine rootlets called **hyphae**. Hyphae feed on timber to produce more spores and the cycle continues. The two most common types of fungi encountered in the construction industry are **dry rot** and **wet rot**.

Dry rot is the most common and damaging fungus. It is off-white in colour and spreads over timber like fluffy cotton wool. Once formed, it produces its own moisture. After the attack, the timber darkens in colour and is left with deep cracks along, and across, the grain. The timber loses its strength and crumbles into a powder. Dry rot is often found in places where the moisture content of timber is between 20% and 35% and where there is a lack of ventilation.

Timber affected by dry rot must be removed and replaced with preserved timber. If brickwork has been affected, then it must be burnt with a blowlamp; dry rot cannot survive beyond a temperature of 40 °C. It is also important to provide adequate ventilation so that the damp conditions never occur in the first place.

Compared with dry rot, wet rot is not such a big problem and usually occurs if the moisture content of timber is above 25%. To make timber really wet, there is usually a structural defect, such as leaking gutters or persistent leakage behind baths. Wet rot causes the timber to become dark brown or black before it splits. Before the affected timber is removed and the adjacent timber treated, it is important to locate the source of moisture and deal with it.

Insect Attack

As with dry rot and wet rot, insects also feed on timber, causing damage which is less serious than fungal damage. Many types of beetles, termites, wood wasps and other insects can damage timber, not only in timber yards but also in buildings. The life cycle of a beetle starts with the female beetle laying eggs in cracks and crevices of bare timber. The eggs hatch into the larvae which feed on the timber, leaving unwanted dust behind. The larvae change into pupae (dormant stage) and, in due course, the pupae change into adult beetles. The adult beetles emerge from the timber through tiny holes known as flight holes, to mate and start the life cycle again. The most common beetles found in the UK are discussed here.

The common furniture beetle is 2–5 mm long and is dark reddish in colour. It attacks both hardwoods and softwoods, particularly their sapwoods. It is most often found in furniture and joinery. The flight holes are circular with a diameter of about 2 mm.

The house longhorn beetle is black and 10–20 mm long. It attacks the sapwood of softwoods only. The flight holes are oval and 5–10 mm in diameter.

The death-watch beetle is brown and 8 mm long. It attacks hardwoods, particularly old oak that has a high moisture content.

The powder post beetle attacks the sapwood of most hardwoods and is found particularly in timber yards and sawmills. Softwoods are not attacked by this beetle. The beetle is reddish-brown, 4–5 mm long and leaves a flight hole of 2 mm (approximately) in diameter.

16.7.4   Preservation

Fungi and insects use timber as a source of food, making it unsuitable for structural purposes. The best way to control this damage is to treat timber with chemicals that are poisonous to fungi and insects, but not to humans. Three types of preservatives are available: creosote substitutes, water-borne preservatives (copper-chromearsenate) and organic solvent preservatives (chlornaphthalenes, metallic naphthanates and pentachlorophenol).

16.7.5   Environmental Implications

The environmental impact of the use of timber is the least of all building materials. Timber is a renewable source, as the forests can be managed to grow trees on a sustained, continuous basis. Trees absorb carbon dioxide from the atmosphere as part of the photosynthesis process. Waste timber, in good condition, can be processed to manufacture products like chipboard.

16.8   Plastics

Plastics are relatively new materials as compared to timber, steel and other construction materials, and they have replaced the traditional materials in several applications. Typical examples where plastics have gained more popularity than the traditional materials are drainpipes, windows, external doors, fascias, soffit boards and cladding.

Plastics contain carbon combined with elements such as hydrogen, oxygen, chlorine and nitrogen. Organic substances containing only the elements carbon and hydrogen are known as hydrocarbons; typical examples are methane, ethane, propane, butane and pentane. The formulae of a selection of hydrocarbons are shown here:

(a)

$$H - \overset{\displaystyle H}{\underset{\displaystyle H}{\overset{|}{\underset{|}{C}}}} - H$$

Methane (CH$_4$)

(b)

$$H - \overset{\displaystyle H}{\underset{\displaystyle H}{\overset{|}{\underset{|}{C}}}} - \overset{\displaystyle H}{\underset{\displaystyle H}{\overset{|}{\underset{|}{C}}}} - \overset{\displaystyle H}{\underset{\displaystyle H}{\overset{|}{\underset{|}{C}}}} - H$$

Propane (C$_3$H$_8$)

(c)

$$H - \overset{\displaystyle H}{\underset{\displaystyle H}{\overset{|}{\underset{|}{C}}}} - \overset{\displaystyle H}{\underset{\displaystyle H}{\overset{|}{\underset{|}{C}}}} - \overset{\displaystyle H}{\underset{\displaystyle H}{\overset{|}{\underset{|}{C}}}} - \overset{\displaystyle H}{\underset{\displaystyle H}{\overset{|}{\underset{|}{C}}}} - \overset{\displaystyle H}{\underset{\displaystyle H}{\overset{|}{\underset{|}{C}}}} - H$$

Pentane (C$_5$H$_{12}$)

As the chain becomes longer, or the molecule becomes larger, the hydrocarbons change from gas to liquid state. Later in the series, the viscosity of the liquid hydrocarbons increases and they become solid. The substitution of other elements in place of one or more hydrogen atom will produce further homologous series. For example, replacing one atom of methane or ethane with a chlorine atom produces methyl chloride or ethyl chloride, respectively.

16.8.1   Raw Materials and Manufacture

Plastics are made from **monomers** which are obtained from the fractional distillation of coal or petroleum. Fractional distillation of petroleum yields many substances: methane, ethane, propane, butane, hexane etc. are some of them. They are all mixed together initially but separated in refineries for use in making different types of plastic. Similarly, between 200 °C and 250 °C, naphthalene and phenol are obtained due to the fractional distillation of coal. Phenol is further used in the production of plastics.

In the next stage of manufacturing, the monomers are subjected to high temperatures and pressures in the presence of catalysts. The monomers are converted into polymers which consist of a large number of monomers. The polymer is compressed into plastic pellets which are further mixed with additives such as pigments, antioxidants, plasticisers, flame retardants, ultraviolet stabilisers etc. to achieve the desired properties in the end product. The mixture is used in the appropriate manufacturing process (extrusion, injection moulding or blow moulding) to produce the desired product.

16.8.2   Classification

Plastics can be classified into two types: thermoplastics and thermosetting plastics.

Thermoplastics
Thermoplastics can be repeatedly melted down by heating, and they harden on cooling without any change in properties. Thermoplastics can be remoulded several times into any desired shape. Typical examples are: polyethylene, polystyrene and PVC.

Thermoplastics have good weathering properties, but may discolour in sunlight. They show creep characteristics and are usually inflammable at high temperatures.

Thermosetting Plastics
Thermosetting plastics harden on the application of heat. Once set in the moulded stage, they cannot be melted and remoulded. Typical examples are: bakelite and urea-formaldehyde.

Thermosetting plastics are hard-wearing, stronger than thermoplastics and resistant to abrasion. The disadvantages are that they are brittle, inflammable and not very flexible.

16.8.3 Properties and Uses

Polyethylene

The scientific name is poly(ethane) but it is more commonly known as polythene, which is ICI's trade name. Polyethylene can be produced by following different methods of polymerisation, and hence its properties vary as well. Two common forms of this polymer are low-density polyethylene (LDPE) and high-density polyethylene (HDPE). In LDPE the chains do not lie close together, resulting in a weaker plastic. It has a lower density (910 to 940 kg/m^3) and a lower melting point (110 °C). LDPE is more flexible than HDPE, and is used in making moisture-barrier films.

In HDPE the molecules lie close together, resulting in a higher density (941 to 970 kg/m^3), a higher melting point (135 °C) than LDPE and higher strength. HDPE is used in manufacturing water pipes and cable insulation. Polyethylene is a poor conductor of heat, having a thermal conductivity between 0.30 and 0.40 W/m K. The coefficient of thermal expansion ranges between 2×10^{-5} and 20×10^{-5} per K.

Both plastics show good resistance to vegetable oils, acids and alkalis.

Polyvinyl Chloride

Polyvinyl chloride (PVC) is a thermoplastic and the most widely used plastic in the construction industry. It is available as plasticised polyvinyl chloride as well as unplasticised polyvinyl chloride (PVCu) and is used in a variety of constructional and non-constructional applications. PVCu is strong (at normal temperatures), durable, lightweight and flame retardant. PVCu is also resistant to dilute acids, dilute alkalis and domestic chemicals, but has poor resistance to aromatic hydrocarbons like naphthalene. Due to these properties, PVCu is extensively used in making window and door frames, pipes and pipe fittings, rainwater gutters, fascias and soffit boards, cladding and conservatories.

PVC is combustible, giving off toxic hydrogen chloride fumes; however, PVCu tends to burn only with difficulty. Typical values of density for PVC and PVCu are 1300 kg/m^3 and 1400 kg/m^3. PVC and PVCu are good thermal insulators with thermal conductivity values of 0.15 W/m K and 0.20 W/m K, respectively. A typical value of the coefficient of thermal expansion is 6×10^{-5} per K.

Pure PVC is a tough material with good resistance to weather and chemicals. Its resistance to heat and light is comparatively poor and, as a result, chlorine is lost when it is exposed to heat and UV light. The loss of chlorine is avoided by adding stabilisers, which are salts of heavy metals like lead and cadmium.

Lead and cadmium compounds used in PVC are toxic and dangerous to the environment. Some cadmium compounds are classified as carcinogens. The use of lead and cadmium stabilisers is safe during the use phase, but contamination of the environment can take place during production and waste treatment. Measures should be taken to eliminate, or to reduce to a minimum, the exposure of workers to these toxic substances.

PVC, which is originally rigid, is made flexible by the addition of substances known as plasticisers and is used in manufacturing upholstery, flooring, roofing membranes and insulation for electrical cables. Naphthalates, especially diethylhexyl phthalate (DEHP), are quite commonly used for this purpose. Some organisations have raised concerns over the toxicity of naphthalates, and some countries have banned their use in children's plastic toys.

Polytetrafluoroethylene (PTFE)

PTFE is a thermoplastic consisting of carbon and fluorine; it is resistant to high temperature, chemical reaction and corrosion, and has a very low coefficient of friction (0.05 to 0.1). These properties make it useful in several applications. It was developed by DuPont in 1938, and marketed as Teflon.

Because of its resistance to high temperature and its non-stick properties, its initial use was to make non-stick kitchen utensils. Due to its resistance to chemicals, it is used for machinery parts that could be susceptible to corrosion, and due to its very low coefficient of friction, it is used in bridge bearings to allow expansion or contraction of bridge decks. PTFE tape is used by plumbers to seal threaded connections, as the tape performs a lubricating action. PTFE is white with a density of 2150 to 2200 kg/m^3. Typical values of thermal conductivity and the coefficient of thermal expansion are 0.25 W/m K and 12×10^{-5} per K.

Melamine Formaldehyde (MF)

Melamine formaldehyde is a thermosetting plastic made from the polymerisation of formaldehyde with melamine. MF is odour free, can be easily coloured and is resistant to heat, light, fire, chemicals and abrasion. MF is used as an adhesive in manufacturing particle board (chipboard) and plywood. Particles of cheap wood, sawdust and paper are glued with melamine formaldehyde and compressed under high pressure to produce particle board. Melamine formaldehyde is also used as a protective covering in laminated boards that are used in furniture, flooring and worktops. The density and the tensile strength are 1400 kg/m^3 and 60 N/mm^2, respectively. The coefficient of linear expansion is 35×10^{-6} per K.

Melamine formaldehyde is affected by alkalis and concentrated acids.

16.9   Glass

Glass is used in several applications in buildings, the main ones being its use in windows, doors and cladding. The main constituents of glass are sand (silica; about 70%), soda, limestone and additives such as oxides of iron, manganese and other metals to improve the physical properties. The raw materials are weighed and mixed with broken glass, which lowers the melting point. The mixture is charged into the furnace, and at around 1550 °C the mixture of raw materials melts. After homogenisation and removal of gas bubbles, the molten glass is allowed to cool slightly and floated onto a mirror-like surface of molten tin to achieve a perfectly flat surface. The thickness of glass is controlled by selecting the rate at which it spreads.

The rigid glass ribbon leaves the molten tin bath at about 600 °C and passes through a cooling tunnel called a **lehr**. The temperature of glass is lowered gradually to 250 °C to remove internal stresses so that it can be cut and worked safely. The glass ribbon is later cut into sheets; Figure 16.8 shows the sequence of operations.

16.9.1   Properties

Density
The density of glass is 2500 kg/m^3.

Thermal Conductivity
Glass is dense and a good conductor of heat; its thermal conductivity is 1.02 W/m K. The U-value of a single-glazed window is approximately 5.6 W/m^2 K, which can be improved by using double-glazed units. The U-value of a double-glazed window, i.e. 2.8 W/m^2 K can be further improved by reducing the heat exchange between the two glass sheets. This can be achieved by using glass with a low-emissivity (low-E) coating, which will reduce the radiated heat transfer between the two glass panes. Heat loss by conduction and convection can be reduced by replacing the air in the cavity with a gas

(for example, argon) that has a lower value of thermal conductivity. Depending on the emissivity coating and gas fill, U-values as low as $1.0\,W/m^2\,K$ can be achieved for double-glazed units.

Strength

Glass in buildings is required to resist loads including wind loads, impact by animals and people and thermal stresses. The compressive strength of glass is very high, far greater than that of some other building materials; therefore, glass does not give any problem in this respect when used in building applications. However, because of microscopic cracks in the surface, the tensile strength of glass is so low that it fails by rapid brittle fracture, and there is no warning of impending failure. The tensile strength of soda-lime glass for practical purposes is 30 to $80\,N/mm^2$. The toughening process will make glass much stronger in tension. Patterns and wires in glass reduce its strength.

Elasticity

Glass is a perfectly elastic material; however, it is fragile and will break without warning if subjected to excessive stress. Young's modulus of glass is $70\,kN/mm^2$.

Linear Expansion

The coefficient of linear expansion is $9 \times 10^{-6}\,m/m\,K$. As the properties of glass are different from the properties of the material in which it is fixed (for example, glass in PVCu), allowance should be made for its expansion/contraction.

Appearance

Ordinary glass is transparent and more or less colourless. The surfaces of float glass are bright and lustrous.

Durability

Glass is a highly durable material; there are only a few factors that can cause any damage. One of the main factors for damaging glass is excessive load. As discussed earlier, the tensile strength of glass is very low; therefore, if the loading causes the glass to bend, there is a chance that it will fail. Acids and alkalis can cause damage to the chemical composition of the glass surface; weak acids can be found in acid rain and alkalis can be found in cement and cement-based products. Alkaline paint removers, if deposited on glass, can also cause damage to the glass surface. To protect the surface of glass from

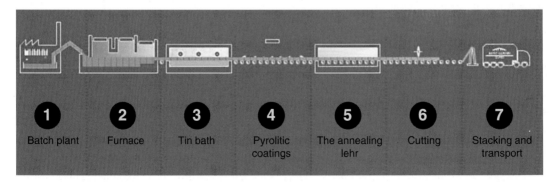

Figure 16.8 Glass manufacturing process. Reproduced by permission of Saint-Gobain, UK. (See the colour plate section for a full-colour version of this image.)

long-term damage, chemicals, if deposited during the construction of a building, should be removed immediately.

Sound Insulation (Section 16.9.2 gives Details on PLANICLEAR and STADIP Glass)

The sound insulation of glass depends on its thickness. For single-glazed windows (Planiclear glass), the sound reduction index (R_w) is 30 dB for 4 mm thick glass. The sound insulation values increase as the thickness increases, for example, for 12 mm thick glass, the sound reduction index is 36 dB. The insulation value could be impaired if there is air leakage around the opening lights. Well-fitted, sealed double-glazed windows are necessary for a superior level of insulation; for a 26.8 mm unit (for example, Planiclear-STADIP Silence glass), the value of R_w is 39 dB.

Fire Resistance

Although non-combustible, ordinary glass breaks and later melts in fire. Glass is a good conductor of heat, and the radiation from glass can ignite combustible materials. The main factors affecting the fire resistance of glass are the type of glass, the thickness, the type of frame and the fixing techniques.

16.9.2   Types of Glass

There is a huge range of glass produced by different glass manufacturers. A detailed discussion is not possible here, but a brief description of four types of glass, produced by Saint-Gobain, UK, is given below. For detailed information on these and other types of glass, the reader should refer to their website.

SGG PLANICLEAR®

PLANICLEAR, a float glass with a lower iron content, reduces the level of absorption, thereby increasing the level of solar gain (g value). PLANICLEAR also contributes to a higher level of light transmission; this helps to increase the level of natural daylight and reduces the need for artificial lighting, creating a more comfortable environment. The lower iron content of the glass also reduces the level of green colouring so that the appearance is clearer and more neutral. It is available in an extensive range of thicknesses suitable for a wide variety of applications such as windows, doors, partitions etc. It can be toughened or laminated for safety benefits or silvered to produce mirrors.

SGG STADIP®

STADIP glass is made up of two or more sheets of glass bonded together with a plastic (polyvinyl butyral -PVB) interlayer. After the plastic has been placed between the glass sheets, the whole assembly is subjected to heat and pressure, resulting in the complete adhesion of the PVB and the glass. In the event of breakage, the glass fragments remain bonded to the plastic interlayer, providing safety. STADIP PROTECT glass, which uses a thicker layer of PVB, can be used for a range of applications such as protection against vandalism and burglary, protection against firearms and protection against explosions. STADIP SILENCE laminated acoustic and safety glass can be used in applications where high acoustic insulation is essential, for example, near city centres, railway lines, motorways etc. without compromising the safety performance.

SGG PLANITHERM®

This is a low-emissivity glass and is very effective in reflecting long-wave heat radiation back into a room, thereby minimising heat loss through a window while also maximising solar heat gain and natural light transmission. A combination of microscopically thin multiple metal oxide layers is applied to high-quality clear float glass using a magnetically enhanced cathodic sputtering process under vacuum conditions. Depending on the composition of these transparent coating layers, several different products can be produced, distinguishable by the thermal performance and other properties.

A double-glazed unit incorporating low-E glass is up to three times more thermally efficient than an ordinary double-glazed unit and offers significantly better thermal insulation and improved comfort. It is also an environmentally friendly solution, given the lower CO_2 emissions associated with reduced energy consumption.

SGG DECORGLASS®

DECORGLASS is a patterned glass which has a texture printed onto its surface. Available in a wide range of patterns and colours, it provides an infinite choice of styles and variations. A piece of clear or tinted molten glass is passed between two rollers, which emboss the pattern into the sheet, providing a highly durable finish.

The range offered in this type of glass is highly flexible and is designed for use in a wide range of applications, where that special touch is required. It will subtly transmit and diffuse light, creating an interior ambiance.

References/Further Reading

1 The Brick Industry Association (2006). *Manufacturing of bricks (Technical note 9)*. Reston (US). Website: www.gobrick.com
2 Dean, Y. (1996). *Materials Technology: Mitchell's series*. Harlow: Longman.
3 Fullick, A. and McDuell, B. (2008). *Edexcel AS Chemistry*. Harlow: Pearson Educational.
4 Kääntee, U., Zevenhoven, R., Backman, R. and Hupa, M. (2002). *Cement Manufacturing Using Alternative Fuels and the Advantages of Process Modelling*. R'2002 Recovery, Recycling, Re-integration, Geneva.
5 Lafarge Cement UK (2011). *Cement in Sustainable Construction*. Birmingham.
6 Lyons, A. R. (1997). *Materials for Architects and Builders – An Introduction*. London: Arnold.
7 Mukherjee, A., Kääntee, U. and Zevenhoven, R. (2001). *The Effects of Switching from Coal to Alternative Fuels on Heavy Metals Emissions from Cement Manufacturing*. Proceedings of the 6th International Conference on the Biogeochemistry of Trace Elements, Canada.
8 Saint-Gobain Glass. Website: www.saint-gobain.com
9 Taylor, G. D. (2000). *Materials in Construction – An Introduction*. Harlow: Longman.

17

Assignments

17.1 Assignments for Level 2 Courses

17.1.1 Assignment No. 1

Task 1

a) Explain, briefly, the effects of gravitational force on the structural elements (walls, floors, roof, beams and columns) of a building.
b) Explain the effect of gravitational forces on the shape of structural elements like walls, columns and beams.
c) Explain how the structural elements of a building may fail as a result of overloading.
d) A 6.0 m long concrete beam that is 200 mm × 400 mm in cross-section is resting on two walls. Calculate the dead load of the beam if the density of concrete is 2400 kg/m^3.
e) Calculate reactions R$_1$ and R$_2$ for the beam shown in Figure 17.1.

17.1.2 Assignment No. 2

Task 1

a) Describe, briefly, what happens to plastic guttering and PVCu windows when the air temperature rises during the day and drops during the night.
b) Describe, briefly, what happens to water when:
   i) Its temperature is lowered to −3 °C.
   ii) Its temperature is raised to 100 °C.
c) Explain the effect of temperature change, in terms of latent heat and sensible heat, on the physical changes in the state of water.
   i) A 5.0 m length of steel tube is heated from 10 °C to 40 °C. Calculate the thermal movement if the coefficient of linear expansion of steel is 0.000012.
   ii) A 4.0 m length of steel tube is cooled from 45 °C to 20 °C. Calculate the thermal movement if the coefficient of linear contraction of steel is 0.000012.

17.1.3 Assignment No. 3

Task 1

Explain how sub-zero temperatures can affect porous materials such as bricks, concrete blocks and roof tiles saturated or partially saturated with water. How are these materials affected by repeated freeze–thaw cycles?

Construction Science and Materials, Second Edition. Surinder Singh Virdi.
© 2017 John Wiley & Sons Ltd. Published 2017 by John Wiley & Sons Ltd.
Companion website: www.wiley.com/go/virdiconstructionscience2e

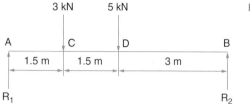

Figure 17.1

Task 2

a) Select two materials and evaluate the effect of forces and temperature changes on their properties.

b) Justify the use of the two materials in the construction of a dwelling house.

17.2 Assignments for Level 3/4 Courses

17.2.1 Assignment No. 1

Introduction

The Midlands Institute of Engineering and Technology has proposed to construct a single-storey conference centre as an extension to the main building. The size of the extension and other details are given in the data sheet. Although the main use of this extension will be for conferences and seminars, it may also be used for other events like lectures, open days etc.

The choice of materials, components and construction techniques should ensure that the new construction conserves energy and creates a comfortable environment in the conference room. The constructional details of the extension are:

    Roof: RC slab

    Ground floor: Solid concrete floor

    Walls: Cavity construction using bricks and dense concrete blocks

Your assignment, as a trainee technician, is to complete the following tasks and present your findings in the form of a written report/PowerPoint presentation. Use tables, images and graphs wherever appropriate.

Task 1

a) Identify and describe the factors (temperature, humidity, ventilation, acoustic environment and visual environment) that could affect human comfort in the conference room.

b) Describe, with diagrams of appropriate instruments, how each factor is measured and state the acceptable values of each identified factor.

Task 2

a) Use the information given in the data sheet and calculate the fabric heat loss and the heat loss due to ventilation from the conference room.

b) Three vehicles, parked in the car park, start at the same time and produce noise levels of 76 dB, 80 dB and 83 dB. Use either the analytical method or the graphical method to calculate the overall noise level.

c) A street lamp (point source) with a luminous intensity of 1200 cd is suspended 6 m above the edge of the road leading to the car park. If the road is 8 m wide, calculate:
 i) The illuminance at point A on the road surface that is directly under the lamp.
 ii) The illuminance at a point, on the other edge of the road, directly opposite point A.

Task 3
a) Explain how the human thermoregulatory system is affected by air temperature, air movement, humidity, clothing and level of activity. Give acceptable values of these factors for the conference room.
b) Discuss how the NC/NR curves and the speech interference criteria may be utilised to analyse noise and achieve acoustic comfort in the conference room.
c) Discuss the reasons why designing for high levels of daylight in the conference room may impact upon levels of visual and thermal comfort. Give acceptable values of illuminance for the conference room.

Data sheet

Plan area of the extension (internal): 20.0 m × 15.0 m

Height of the building: 4.5 m (floor to ceiling)

Number of 1.5 m × 1.2 m high double-glazed windows: 12

Number of 2.0 m × 2.0 m high doors: 3

Internal temperature and humidity: 21°C and 60%, respectively

External temperature and humidity: 0°C and 100%, respectively

Number of air changes: 2 per hour

U – values $\left(\text{W/m}^2\text{K}\right)$

Floor: 0.50; Roof: 0.50; Windows: 1.70; Doors: 2.0

(Refer to Chapter 8 for any other information)

17.2.2 Assignment No. 2

Task 1
a) The conference centre will be affected by different types of forces when the construction is over. Explain the following forces:
 i) Point load and uniformly distributed load.
 ii) Dead load, imposed load and wind load.
 iii) Concurrent, non-concurrent and coplanar forces.
b) Explain and differentiate between compressive, tensile, shear and bending stresses.
c) Explain how the above stresses change the shape of elements/members they act upon.
d) Describe, briefly, how excessive stresses may lead to the failure of elements/members.

Task 2
a) Explain how the forces acting on beams, joists, floors etc. can cause compressive stress, tensile stress and shear stress at the same time.

b) Explain how forces can cause the following in simply supported beams and cantilevers:
   i) Positive and/or negative bending.
   ii) Positive and negative shear.

Task 3
a) Find reactions for the beams shown in Figure 17.2.
b) A metal rod of 10 mm diameter had an initial length of 1.0 m. After a tensile force of 12 kN was applied to the rod, its length increased to 1.002 m. Calculate the modulus of elasticity of the metal.
c) Figure 17.3 shows two rods, A and B, supporting a load of 400 N. Find the nature and magnitude of forces in the two rods.

17.2.3 Assignment No. 3

Introduction
Several materials will be used in the construction of the conference room. These are: cement, concrete, bricks, concrete blocks, timber, plastic, steel, gypsum plaster, glass, copper and others.

(a)   Figure 17.2

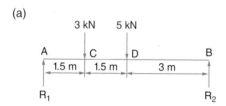

(b)

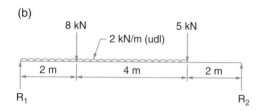

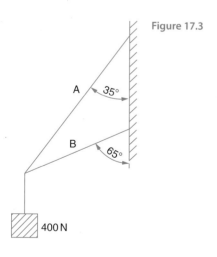

Figure 17.3

Select any THREE that are relevant to your course/profession and complete the following tasks. You can present your findings either in the form of a word-processed report or a PowerPoint presentation. Use tables, images and graphs wherever appropriate.

Task 1

a) Describe the process for producing/manufacturing your selected materials. In each case, comment on the amount of energy consumed in the production/manufacturing process.
b) Describe in detail the main properties (for example, strength, elasticity, thermal movement etc.) of the selected materials.
c) Describe the processes that may cause the deterioration/failure of the selected materials, after their use in the conference centre.
d) Explain the preventative and remedial techniques that may be used to prevent deterioration of construction materials, and compare their effectiveness.

Task 2

Justify the use of your selected materials in the construction of the conference centre.

Appendix 1

Formulae for Example 8.2

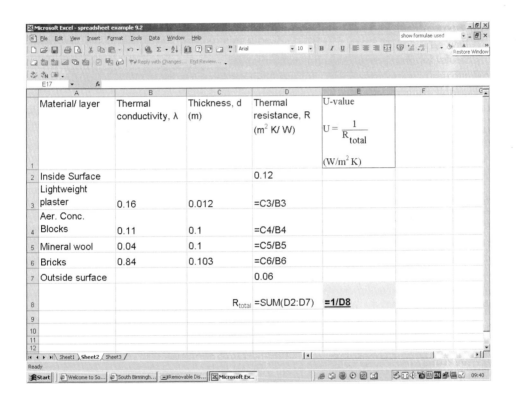

Example 8.2 This spreadsheet shows how the formulae are entered to calculate the thermal resistance of plaster, bricks and other materials. The syntax for adding a number of values is also shown.

Construction Science and Materials, Second Edition. Surinder Singh Virdi.
© 2017 John Wiley & Sons Ltd. Published 2017 by John Wiley & Sons Ltd.
Companion website: www.wiley.com/go/virdiconstructionscience2e

Appendix 2

Solutions for Example 13.10

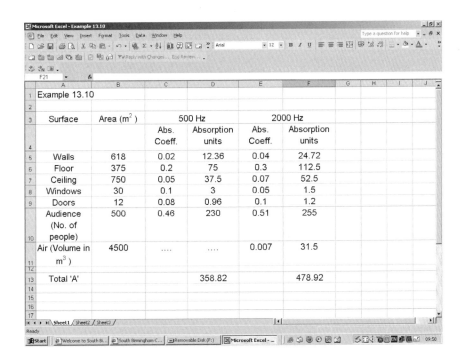

Surface	Area (m^2)	500 Hz		2000 Hz	
		Abs. Coeff.	Absorption units	Abs. Coeff.	Absorption units
Walls	618	0.02	12.36	0.04	24.72
Floor	375	0.2	75	0.3	112.5
Ceiling	750	0.05	37.5	0.07	52.5
Windows	30	0.1	3	0.05	1.5
Doors	12	0.08	0.96	0.1	1.2
Audience (No. of people)	500	0.46	230	0.51	255
Air (Volume in m^3)	4500	....	....	0.007	31.5
Total 'A'			358.82		478.92

Construction Science and Materials, Second Edition. Surinder Singh Virdi.
© 2017 John Wiley & Sons Ltd. Published 2017 by John Wiley & Sons Ltd.
Companion website: www.wiley.com/go/virdiconstructionscience2e

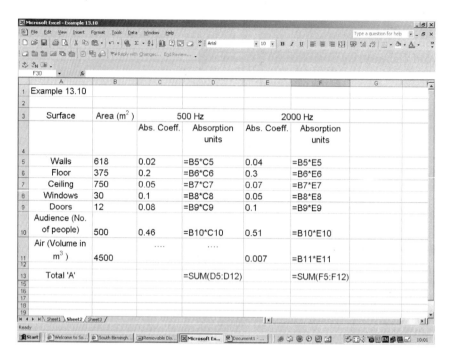

Surface	Area (m²)	500 Hz		2000 Hz	
		Abs. Coeff.	Absorption units	Abs. Coeff.	Absorption units
Walls	618	0.02	=B5*C5	0.04	=B5*E5
Floor	375	0.2	=B6*C6	0.3	=B6*E6
Ceiling	750	0.05	=B7*C7	0.07	=B7*E7
Windows	30	0.1	=B8*C8	0.05	=B8*E8
Doors	12	0.08	=B9*C9	0.1	=B9*E9
Audience (No. of people)	500	0.46	=B10*C10	0.51	=B10*E10
Air (Volume in m³)	4500	....	....	0.007	=B11*E11
Total 'A'			=SUM(D5:D12)		=SUM(F5:F12)

Cell F30 (Example 13.10)

Appendix 3

Answers to Exercises

Answers to Exercise 1.1

1) **A** 98.1 N   **B** 588.9 N
2) 27.6 kg
3) Mass = 3024 kg; Weight = 29665.44 N
4) 12 m/s^2
5) 188.89 N
6) **A** 2707.56 N   **B** 40613.4 J
7)

Height (m)	$v^2 = 2as + u^2$	Kinetic energy (KE)	Potential energy (PE)	Total energy = PE + KE
9	0	0 J	706320 J	706320 J
6	58.86	235440	470880	706320
3	117.72	470880	235440	706320
0	176.58	706320	0	706320

8) **A** 245250 J;   **B** 20437.5 W

Answers to Exercise 2.1

1) See Section 2.3
2) See Table 2.3
3) **A** $CaCl_2$ **B** $CaSO_4$ **C** $MgCl_2$
4) **A** and   **B** See Section 2.5.1
5) **A** See Section 2.5.2
   B NaOH
   C Chemical reaction between the two occurs which produces hydrogen gas and a salt. The strength and the texture of marble are affected.
   D Marble + dilute sulphuric acid → calcium sulphate + water + carbon dioxide gas
6) **A** Sodium + dilute hydrochloric acid → sodium chloride + hydrogen gas
   B Sodium + dilute sulphuric acid → sodium sulphate + hydrogen gas
   C Iron + dilute hydrochloric acid → ferrous chloride + hydrogen gas

Construction Science and Materials, Second Edition. Surinder Singh Virdi.
© 2017 John Wiley & Sons Ltd. Published 2017 by John Wiley & Sons Ltd.
Companion website: www.wiley.com/go/virdiconstructionscience2e

    D Iron + dilute sulphuric acid → ferrous sulphate + hydrogen gas

    E Aluminium + hydrochloric acid → aluminium chloride + hydrogen gas

    F Sulphuric acid + sodium hydroxide → sodium sulphate + water

Answers to Exercise 4.1

1) 400 V
2) p.d. across 4 Ω resistor = 16 V; p.d. across 8 Ω resistor = 32 V Supply voltage = 48 V
3) 2 Ω
4) **A** 3 Ω **B** 4 A   **C** Current through 4 Ω resistor = 3 A; Current through 12 Ω resistor = 1 A
5) **A** 6 Ω **B** Current in R_1 = 2 A; Current in R_2 = 1.5 A; Current in R_3 = 0.5 A
6) **A** 55   **B** 4.78 A

Answers to Exercise 7.1

1) **A** 283 K   **B** 303 K   **C** 223 K
2) **A** 22°C   **B** 42°C   **C** 70°C
3) 22.8×10^{-6} per°C
4) 1.8 mm

Answers to Exercise 8.1

1) 1.81 W/m^{2}K
2) **A** 2.16 W/m^{2}K   **B** 78 mm
3) **A** 0.68 W/m^{2}K   **B** 0.39 W/m^{2}K
4) 2835.73 W
5) See Figure A8.1-1

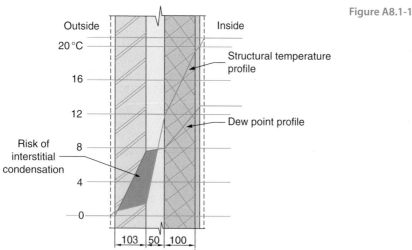

Figure A8.1-1

Answers to Exercise 9.1

1) 0.04 kN/mm^2
2) 0.00133 or 0.133%
3) 95.49 kN/mm^2
4) See Figure A9.1-1
5) 1.67 mm
6) **A** 25.75 mm   **B** 2.13 mm

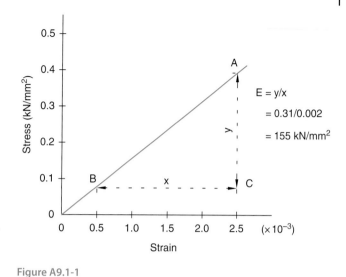

$E = y/x$

$= 0.31/0.002$

$= 155 \text{ kN/mm}^2$

Figure A9.1-1

Answers to Exercise 10.1

1) **A** 24.5 kNm   **B** 11.25 kNm
2) **A** About B: CW moment = ACW moment = 19.25 kNm About A: CW moment = ACW moment = 8.25 kNm
   B About B: CW moment = ACW moment = 21 kNm About A: CW moment = ACW moment = 28 kNm
3) **A** $R_1 = 3.43$ kN, $R_2 = 4.57$ kN
   B $R_1 = 4.75$ kN, $R_2 = 3.25$ kN
   C $R_1 = 6.67$ kN, $R_2 = 3.33$ kN
   D $R_1 = 14.75$ kN, $R_2 = 10.25$ kN
4) **A** See Figure A10.1-1
   B See Figure A10.1-2
   C See Figure A10.1-3
   D See Figure A10.1-4
5) Force in string A = 19.4 N (Tension) Force in string B = 23.0 N (Tension)
6) **A** Force in rod A = 600 N (Tension) Force in rod B = 412 N (Compression)
   B Force in rod A = 205 N (Tension) Force in rod B = 300 N (Tension)
   C Force in rod A = 728 N (Tension) Force in rod B = 460 N (Compression)
7) **A** See Figure A10.1-5
   B See Figure A10.1-6

8 kN

A

4 m

3 m

B

$R_1 = 3.43$ kN

(a)

$R_2 = 4.57$ kN

3.43 kN

+

(b) Shear force
diagram

−

4.75 kN

+ 13.72 kNm

(c) Bending moment diagram

(a)

3 kN

5 kN

A

B

$R_1 = 4.75$ kN

$R_2 = 3.25$ kN

4.75 kN

+

3 kN

(b) Shear force
diagram

5 kN

−

3.25 kN

+

7.13 kN

9.75 kN

(c) Bending moment diagram

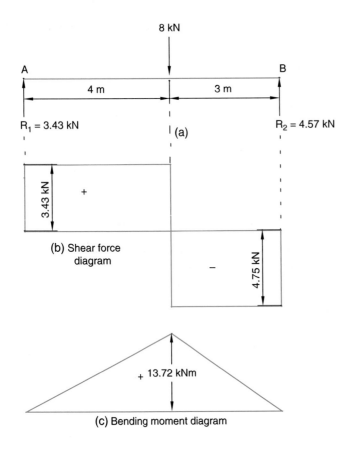

Figure A10.1-3

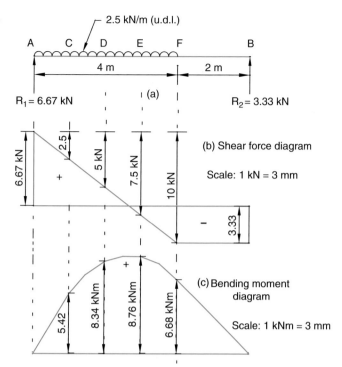

2.5 kN/m (u.d.l.)

A C D E F B

4 m   2 m

$R_1 = 6.67$ kN   (a) $R_2 = 3.33$ kN

6.67 kN   2.5   5 kN   7.5 kN   10 kN

+

(b) Shear force diagram

Scale: 1 kN = 3 mm

−   3.33

5.42   8.34 kNm   8.76 kNm   6.68 kNm

+

(c) Bending moment diagram

Scale: 1 kNm = 3 mm

Figure A10.1-4

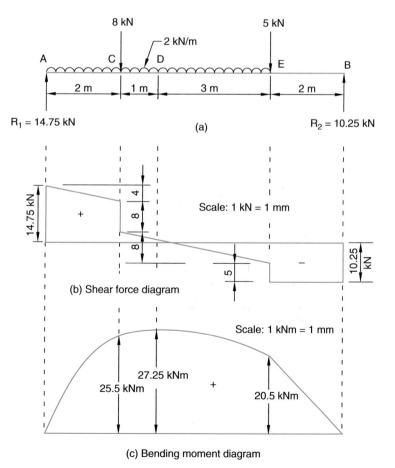

8 kN   5 kN

2 kN/m

A   C   D   E   B

2 m   1 m   3 m   2 m

$R_1 = 14.75$ kN   (a) $R_2 = 10.25$ kN

14.75 kN   4   8   8

+

Scale: 1 kN = 1 mm

−   10.25 kN

5

(b) Shear force diagram

Scale: 1 kNm = 1 mm

27.25 kNm

25.5 kNm   +   20.5 kNm

(c) Bending moment diagram

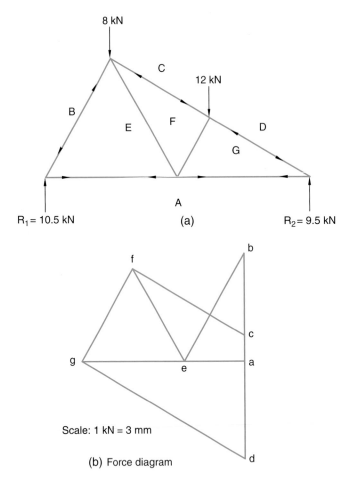

(a)

R₁= 10.5 kN

R₂= 9.5 kN

Scale: 1 kN = 3 mm

(b) Force diagram

$R_1 \times 8 = 8 \times 6 + 12 \times 3$;

$R_1 = 10.5$ kN; $R_2 = 9.5$ kN

Member	Compression (kN)	Tension (kN)
BE	12.1	
CF	13.0	
FE		10.4
EA		6.1
DG	19.0	
GF	10.4	
GA		16.5

Figure A10.1-5

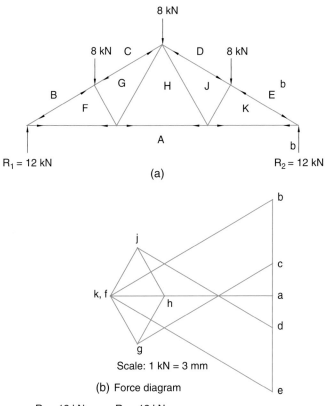

(a)

(b) Force diagram

Scale: 1 kN = 3 mm

R_1 = 12 kN; R_2 = 12 kN

Member	Compression (kN)	Tension (kN)
BF, EK	24.0	
CG, DJ	20.0	
FA, KA		20.8
GF, KJ	6.9	
HG, JH		6.9
HA		13.9

Figure A10.1-6

Answers to Exercise 12.1

1) **A** Base: 88.290 kN; 3 m side: 33.109 kN; 2 m side: 22.073 kN
   B 0.5 m
2) **A** 847.584 MN   **B** 8 m
3) **A** 3310.875 kN   **B** 2.1 m
4) 7691.04 N

5) 1.273 m/s and 2.264 m/s
6) 313.67 kN/m^2
7) 0.0334 m^3/s
8) **A** 2.816 m/s   **B** 4.747 m/s

Answers to Exercise 13.1

1) 345.6 m/s
2) v = 333.33 m/s; λ = 0.651 m
3) 5073 m/s
4) 56.532 dB
5) 133.06 dB
6) 99.55 dB
7) 105.4 dB (approximate); 105.34 (from analytical method)
8) 27.96 dB
9) **A** 3.78 s (at 500 Hz); 2.57 s (at 2000 Hz)   **B** 1.34 s   **C** 1.67 s
10) 4.34 s (at 500 Hz); 2.86 s (at 2000 Hz)

Answers to Exercise 14.1

1) 3141.6 lx
2) **A** 33.3 lx   **B** 7.2 lx
3) 14.36 lx
4) 1.58
5) UF = 0.65; MF = 0.65
6) 24 luminaires (610 × 600 mm recessed modular luminaries, Quattro T body − 2 × 34 W TC-L lamps)
7) 24 luminaires
8) 125 lx

Index

Construction Science and Materials, Second Edition. Surinder Singh Virdi.
© 2017 John Wiley & Sons Ltd. Published 2017 by John Wiley & Sons Ltd.
Companion website: www.wiley.com/go/virdiconstructionscience2e